Manufacturing Technology Research

www.novapublishers.com

Manufacturing Technology Research

Textile Insights and Future Aspects
Gamze D. Tetik, PhD (Editor)
Gizem Celep, PhD (Editor)
Fulya Yilmaz, PhD (Editor)
2024. ISBN: 979-8-89113-289-4 (Hardcover)
2024. ISBN: 979-8-89113-354-9 (eBook)

Computer-Aided Design: Advances in Research and Applications
Dimitrios Tzetzis, PhD (Editor)
Panagiotis Kyratsis, PhD (Editor)
2023. ISBN: 979-8-88697-609-0 (Softcover)
2023. ISBN: 979-8-89113-147-7 (EPUB)

Additive Manufacturing and Reverse Engineering: Research and Manufacturing of Complex Parts
Betim Shabani (Author)
Vladimir Dukovski (Author)
2021. ISBN: 978-1-53619-718-1 (eBook)

Industry 4.0: Principles, Effects and Challenges
Yilmaz Uygun (Editor)
2020. ISBN: 978-1-53618-331-3 (Hardcover)
2020. ISBN: 978-1-53618-423-5 (eBook)

Goat Milk Chemistry and Its Product Manufacturing Technology
Mingruo Guo (Editor)
2020. ISBN: 978-1-53617-299-7 (Hardcover)
2020. ISBN: 978-1-53617-300-0 (eBook)

More information about this series can be found at
https://novapublishers.com/product-category/series/manufacturing-technology-research/

Jose D. Phillips
Editor

Injection Molding

Advances in Research and Applications

Library of Congress Cataloging-in-Publication Data

ISBN: 979-8-89113-429-4

Published by Nova Science Publishers, Inc. † New York

Contents

Preface

This book is comprised of six chapters on injection molding. In Chapter One the main objective of the study is to recycle post-consumer high-density polyethylene (HDPE) and reuse it in fast-moving consumer goods (FMCG) products. Chapter Two aims to compare the advantages and disadvantages of injection molding and 3D printing and convey the latest developments in these methods. Chapter Three presents an overview of bioplastics' characteristics, and processing in injection molding. Chapter Four studies the effect of Moringa Oleifera leaves on thermo-mechanical properties of low-density polyethylene. Chapter Five explains issues such as the advantages and usability of natural fiber reinforcement polymer composites in terms of environmental sustainability. In the last chapter, different rubber processing methods are explained. Also, the basic concept of rubber injection molding and the factors affecting this process are discussed with relevant literature.

Chapter 1

The Recycling of Post-Consumer High-Density Polyethylene and Reuse in Fast-Moving Consumer Goods Products

Raza Muhammad Khan, ME
and Asim Mushtaq*, PhD
Department Polymer and Petrochemical Engineering,
NED University of Engineering and Technology, Karachi, Pakistan

Abstract

The main objective of the study is to recycle post-consumer high-density polyethylene (HDPE) and reuse it in fast-moving consumer goods (FMCG) products. The main focus of the research is to use post-consumer HDPE waste in cosmetic/shampoo bottles after recycling them. The recycled HDPE is used with some composition of virgin HDPE. Recycle material usually show low resistance against the air that is used to blow the material so there is a chance that air might be blown through the parison without expanding it into the shape of the bottle during the manufacturing of bottles. To avoid this problem proper recycling method must be appropriate to get the material of a uniform composition. The methodology which is using for recycling HDPE is mechanical recycling which has many merits including the important one of segregation. The steps in mechanical recycling include sorting, washing, regrinding, and reprocessing material. By using a twin screw extruder all the grind material is converted into pellets. A hot molding press is used to convert these pellets into different specimens. After pelletization and manufacturing of specimens different rheological and

* Corresponding Author's Email: engrasimmushtaq@yahoo.com.

In: Injection Molding
Editor: Jose D. Phillips
ISBN: 979-8-89113-429-4

mechanical properties of the material are measured. The mechanical tests include the tensile test, flexural test, and Izod / Charpy impact test. The flow behavior of the material is measured by the melt flow index test. All these tests would be done with different compositions of virgin and recycled HDPE. This research would determine the best suitable composition for the product that enhances the service life of the finished product.

Graphical Abstract

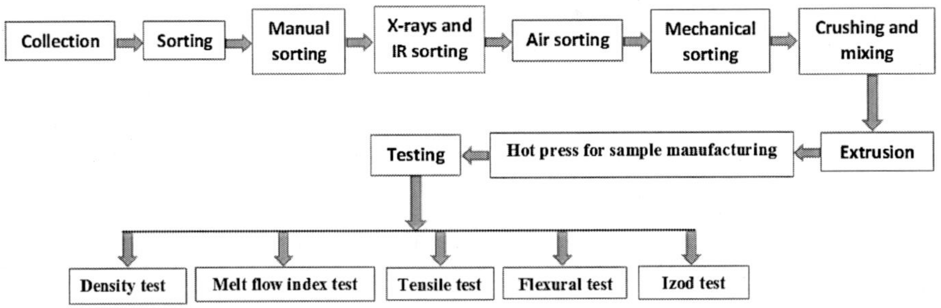

Keywords: high-density polyethylene (HDPE), mechanical recycling, segregation, rheological properties, flexural strength

List of Abbreviation

E-GMA	Ethylene-glycidyl Methacrylate
ESCR	Environmental Stress Crack Resistance
EU	Europe
FFF	Fused Filament Fabrication
FMCG	Fast Manufacturing Consumer Goods
HDPE	High-Density Polyethylene
LCA	Life-Cycle Assessment
LDPE	Low-Density Polyethylene
LLDPE	Linear Low-Density Polyethylene
LLDPE	Linear-Low Density Polyethylene
MFI	Melt Flow Index
NIR	Near Infrared
PCR	Post-Consumer Recycled

PEG	Poly Ethylene Glycol
PET	Poly Ethylene Terephthalate
PET	Polyethylene Terephthalate
PP	Poly Propylene
r-HDPE	Recycled High-Density Polyethylene
R-PET	Recycled poly (Ethylene Terephthalate)
SEBS	Styrene-Ethylene/Butyldiene-Styrene
V-HDPE	Virgin High-Density Polyethylene
VOCs	Volatile Organic Compounds

Introduction

Plastic materials which are made up of polymeric materials have played a vital role in the modern era because as know that they are used in our daily life, and they have very diversified applications that's why they have a very huge demand from consumer sectors. They are cheap compared to other materials which are made up of steel or iron. Due to the increase in demand for plastic materials, its annual production meets about 370 million tons worldwide. Many plastic products have a very short life cycle it is estimated that 40% of those plastic products have a service life of less than one month. Those plastic products which have a very short lifespan place an important contribution to the national market because ultimately, they are sent to landfill sites where they are burned and create pollution. Unfortunately, Pakistan doesn't have any solution to consume those plastic products which have a short life span, or which cannot be recycled or reused. The polymeric materials which are sent to the landfill site include all grades of polyethylene, polystyrene, polyvinyl chloride, and polypropylene. They are the foremost represented in any plastic waste stream. The plastic waste generation in Pakistan is increasing day by day and its inadequate disposal has been a great concern for the government and researchers. The management should adopt several measures one of which is recycling. Everyone should adopt the policy of the 3 R's which are reduced, reuse, and recycle (Williams and Rangel-Buitrago, 2022; Zhu et al., 2022).

Recycling plays a vital role in a country's economy as it generates the outcome for the people as well as helps in the development of countries economy it also contributes to the conservation of the environment. Plastic waste recycling not only helps to reduce the waste from the environment but also contributes to the reduced consumption of energy because it does not use virgin polymer. It uses recycled polymeric waste or in some specific composition. In Pakistan, polymeric waste recycling still faces some bottlenecks because the machinery uses for recycling is inefficient other than that the import duties on the machines are so much higher than industries are not able to compete with the international market. The startups which recycle polymeric waste material have the issue that the product they manufacture has a higher cost than its selling cost. To overcome all these issues the government should help such small startups by subsidizing these types of machinery. You can make the process efficient by following these necessary steps: The proper segregation of polymeric waste and the washing of this material for the removal of fats, food waste, labels, papers, and contaminants in general. All these steps will surely help to improve the quality of the products to be obtained. After washing all polymeric waste the wastewater must be properly treated so that there is not only change in the sort of environmental pollution (Gong et al., 2020; Zhu et al., 2022). Normally recycle material is not used in the same application. This study tends to use recycled material in the same application. The specific ratio of PCR material and virgin material is set by focusing on its application which is cosmetic bottles. This thing makes the project unique from other studies that have already been done (Juan et al., 2020; Sitadewi et al., 2021).

Nowadays plastic pollution is a burning issue, and many nonprofit organizations and governments are working to deal with this matter. This study aims to recycle high-density polyethylene waste and reuse it in FMCG products. This initiative will reduce the debris from landfill sites as well as reduce the pollution from the environment. The recycling of waste encourages the concept of a circular economy and creates jobs for people. This research aim is associated with Unilever Pakistan whose pledge is to recycle all packaging waste by 2025. The process of recycling starts with the collection of HDPE waste from different vendors of Unilever Pakistan. The recycling is done via an EREMA machine followed by sorting, crushing, mixing, extrusion, and sampling processes. The recycled material is physically mixed with virgin HDPE in different ratios. The mixed material is then run into a twin-screw extruder for melt mixing and pelletization. The processing parameter of the twin screw extruder must be carefully set to avoid the thermal degradation of waste material. After extrusion, the specimens are manufactured on a hot press by using a different mold for mechanical testing.

Plastic waste contains a variety of polymers. For recycling, perfect sorting is required for the product's durability, strength, and ease of process. Making flakes/granules/pallets after sorting and ready to use, these as raw materials for the recycling process. According to the study EU, the primary data was obtained from sorting plants and from the recycling of plastic packaging waste to determine the impurities present in waste after sorting, their process efficiency, and material flows. The rate of recycling waste increases when advanced and good technologies are used. The study also suggests that improving product design can significantly lessen the prevalence of impurities and pollutants in input waste, which have a significant impact on recovery rates and the creation of markets for lower-value fractions, such as polypropylene. Almost 60% of waste recycling capacity would be achieved in 2035. Improvising the facility of sorting increases the efficiency of recycling so maximizing the quality and minimizing the rejections. Material selection according to the demand from the waste is important for the benefits of the process and circular economy as well. Recycling plants received the bales of polymer of required demand and flakes or granules or pellets are formed which would be used as the raw material for consumer goods production (Juan et al., 2020; Rosenboom et al., 2022).

Several factors influenced the procedure of post-consumer packaging recycling. Product design and the collecting system also be an effect on it. The odor, color, design, viscosity, and heavy metal content were affected as well. The study is about the purity of collection materials and primary data on recycling performed in the EU. The presence of impurities highly affects the cost, the revenue, and the investment too. Targeting the market requirement product from the waste in a condition of making a consumer good with acceptable and usable properties. The advancement of the recycling industry of plastics necessitates modifications to the recycling system, from collection to expansion. Separating all of the different polymers that enter the recycling stream is one of the most difficult tasks facing the plastics recycling business today. HDPE is one of the most popular and recycled polymers on the market today. A major worry is a contamination from polypropylene in the recovered HDPE stream. Additionally, HDPE which has been designed for recycling is more effective at promoting circularity. Re-extrusion of shredded, recovered plastics during mechanical recycling adds heat and shear to the process. The features of the plastic might be destroyed during this demanding operation. With their greater ESCR and creep resistance, modern HDPE resins allow for the incorporation of higher percentages of degraded recycled material without compromising the functionality of the final product (Colwill et al., 2017; Correa et al., 2022).

Reprocessing of the virgin-recycled HDPE blends exhibits various behaviors. The investigation had been performed by using the blend of virgin and recycled HDPE, and major properties showed affected. Three cycles of processing and reform of virgin-recycled HDPE take place and the properties show degradation in the third reprocessing. On prepared virgin-recycled blends, effects can be measured through different testing methods. Evaluate the properties of blends through the tensile test and MFI test. The result shows that the mechanical characteristics of the mixture decreased by increasing the composition of recycled content in a blend. A reduction was seen after each cycle as compared to the first reprocessing. The viscosity of a virgin-recycled HDPE blend marginally decreases as the recycled percentage increases. The rheological behavior was unaffected by the initial three reprocessing cycles. The material has degraded more quickly as a result of accelerated aging and reprocessing together than by each process alone (Gall et al., 2020; Siracusa and Blanco, 2020).

Melt flow index (MFI) and tensile tests were used in the current experiment to assess the impact of recycled HDPE content, reprocessing cycles, and accelerated aging on the characteristics of the produced blends. According to the rheological data, the quantity recycled, and the reprocessing cycle had a considerable impact on the viscosity values. Particularly, large levels of recovered HDPE and up to three reprocessing processes had little to no effect on the materials' viscosity values. On the other hand, the reprocessing cycle resulted in a more significant drop in viscosity when the least amount of recycled HDPE was added. The acceleration of the diffusion process effectively served as the primary indicator of the deterioration of the submerged material. The shift in molecular weight distribution brought on by the impact of reprocessing cycles was utilized to explain this (Gall et al., 2020; Zhu et al., 2022).

High-density polyethylene (HDPE) degradation can pose a serious challenge for recyclers. Even though HDPE is one of the plastics that is recycled the most frequently and can be separated using a variety of techniques, recycling this polymer is not always as effective as it could be. To address performance and quality degradation and satisfy the high demand for this material, the processing can be adjusted. Any existing wear and tear from prior use may be made worse by the heat and shear generated during the HDPE recycling process. The performance and quality of the recycled may be adversely affected by these degradation-related changes to the resin structure, such as the introduction of gels and specks, the onset of side reactions that change the polymer's color, or variations in melt flow. Acidic species and other contaminants from the recycle stream can start the degradation process, making them "pro-degrades." They may be introduced from the recycling stream, or they may be present as a result of catalyst residues left over from the production of polymers, which led to side reactions that produced acids. If an antacid is not also present, acids weaken the HDPE polymer chains' structural integrity and reduce the potency of antioxidants (Bauer et al., 2022; Morgan et al., 2022).

Post-consumer recycled (PCR) content is in high demand worldwide thanks to major brands' commitments to sustainability and new laws that support circularity by promoting plastic recycling. In particular, the market for recycled HDPE in natural colors has grown in 2021, primarily as a result of rising demand from companies that manufacture consumer packaged goods (CPG). Recyclers must preserve post-consumer HDPE characteristics, including color, throughout the recycling process to supply the market. Stabilizing the PCR content before exposing it to heat and shear stresses is a tried-and-true technique. The HDPE PCR content can be chemically stabilized to prevent yellowing and keep the resin's original color, maintaining the virgin resin's physical characteristics, such as impact strength. Keep the melt flow properties required for effective processing and prevent the formation of gels and other artifacts to guarantee the homogeneity of the resin structure. Numerous products, including non-food bottles, pipes, pallets, and decking, can be made from recycled HDPE. Although HDPE is comparatively simple to recycle, heat, shear, and contamination during the recycling process can harm the material's color and functionality. Chemical stabilizers, in particular stabilizer blends, can be added to the melt during recycling to prevent product deterioration, improve melt flow consistency, and preserve the desired natural color (Shanker et al., 2022; Zhu et al., 2022).

Due to their lightweight nature, HDPE polymers are increasingly frequently used, and their products are chosen over metals, simple manufacturing process, strength, and toughness. Because of this, manufacturers favor HDPE polymers as their preferred choice. However, one disadvantage is that this polymer takes a while to degrade and has an impact on the environment when disposed of as garbage. For the production of HDPE pipes and fittings, almost all industries solely employ virgin imported materials, which raise manufacturing costs and require more virgin material extraction. The goal of this project is to use recycling to solve this issue. In this study, the effects of virgin and recycled HDPE's composition on the mechanical properties of HDPE pipe fitting are being looked at. Different weight percentages of virgin (pure) HDPE were mixed with recycled HDPE. According to the experimental findings, virgin contents up to 30% by mass and recycled contents up to 70% by mass both met the standards. By utilizing the scrap contents, this ideal ratio will enable AG pipe fitting companies to significantly reduce annual costs (Nikiema and Asiedu, 2022; Siracusa and Blanco, 2020).

Recycled HDPE of high quality can be obtained by effective recycling which is directly dependent upon the capability of sorting. Sorting to a high purity means that the plastic can be re-used and retain its properties. In modern recycling plants, the use of Near Infra-Red (NIR) technology is very common. NIR detects the kind of polymer by its reflectivity of it and distinguishes between their wavelengths signatures to separate them optical detection systems are also used to segregate different colors. An organization named prism has tested fluorescent markers that provide a distinct signal to the sorter. A marker can be incorporated into a plastic end product by locking it with printing ink on packaging labels. The priority of this project was to recover food-grade polymer, additionally, these markers also help to recognize black plastic (invisible to NIR). This technology can be used in line with NIR, to differentiate between polymers. After testing it was found that these fluorescent markers help to produce 96% for PET but for food-grade HDPE the standards are higher. Passing it twice the detection system yields 99.6% purity which is above the required 99% for food-grade HDPE. Carbon black which is usually used in pigments in packaging is invisible to NIR one replacement for it is NIR-detectable pigments. Unilever one of the top FMCG suppliers has tested two brands and started producing bottles using this pigment which means tons of plastic bottles can be recycled more efficiently. Another concept of digital watermarks can also be adopted in which the codes are imprinted or engraved which cannot be seen by the naked eye but can be recognized by the camera sorting line uses. Along with recognizing material, the code can carry additional information like how many times it has been recycled. These codes can be stamped on plastics or incorporated into labels. A Dutch company used this technique and claims 98% accuracy using. Recycling can only be done industrially without using any additives if the process of sorting is efficient, the core part of recycling revolves around effective sorting (Boesen et al., 2019; Garofalo et al., 2019). Color pigments and master batches are used to give different colors to plastics. These pigments and master batches are usually made up of based material and normally do not affect the properties of the material.

Important considerations in product design and production include the removal effectiveness of water-based separation systems and the general compatibility of any attachments on the bottles with HDPE resin. It is strongly advised that product designers, manufacturers, and brand owners think about ways to lower the possible dangers of contamination by trying to keep the adhesive and surface covering on the containers to a minimum and avoiding or using glue as little as possible for attachment. If adhesive is used, it should be used sparingly, instead of pressure-sensitive or thermoset polyurethane adhesives, use water- or alkali-soluble adhesives, use water-based glues because they may be readily diluted and removed during the washing process, and avoid utilizing glues, such as pressure-sensitive adhesives, that cannot be melted by hot caustic (Nikiema and Asiedu, 2022; Zhu et al., 2022).

Several issues with recycling HDPE can be brought on by packaging closures and seals. The design and selection of materials for closures and seals should, whenever possible, follow the guidelines. Colored HDPE which is used to manufacture caps of bottles does not cause any significant degradation. The only thing that should be taken into consideration is the off coloring of the produced granules. LDPE which is used in labels can be scrapped off or can be separated using the difference in specific gravity during washing. EVA Closure, the density of EVA is lighter than water, so it is hard to separate them. But it can be recycled along with HDPE to use in low-grade applications. PVC seals are hard to take off from bottles and they will form black specks when recycled with HDPE so it is contaminated because they degenerate into carbon. Aluminum closures are difficult to recycle because it remains on bottles even after the process. It increases the time for recycling and also degrades the quality of the material. Water-soluble inks shouldn't be utilized since they cause the wash water to bleed and get colored, coloring the HDPE flakes in the process (Williams and Rangel-Buitrago, 2022; Zhu et al., 2022).

The most used method for processing HDPE is extrusion, which creates objects with a fixed cross-sectional profile. HDPE pellets are put into the barrel of the extruder, where they are crushed and conducted heat. The desired shape of the finished product is subsequently created by pressing the molten plastic which passes from a die at a certain pressure. High-density polyethylene can be used for making other products after melting, and have an ability because of thermoplastic, but the degree to which the thermomechanical properties and the deterioration of the recycled components are affected by HDPE's structure and processing circumstances. Both the initial processing and subsequent recycling and reprocessing activities result in degradation. Generally speaking, under favorable reprocessing conditions, the recycled material's characteristics are comparable to those of virgin HDPE. The study's sole goal is to evaluate the extruded recycled HDPE hollow parts' mechanical performance. The qualities of the recycled HDPE's flexural yield and ultimate strength turned out to outperform those made of non-recycled HDPE. When compared to non-recycled HDPE, recycled HDPE has a lower tensile yield strength and elastic modulus. Tensile ultimate, compressive yield, and compressive ultimate strength do not significantly change between virgin and r-HDPE. Recycled HDPE has better flexural yield and flexural ultimate strength characteristics as compared to virgin HDPE. Overall, recycled HDPE replaces virgin HDPE, it offers good alternatives for both effective cost production and protection of the environment in light of the aforementioned factors. The investigation that has been mentioned so far was only able to compare r- HDPE and virgin HDPE, but it is expanded upon by mixing various proportions of r-HDPE and virgin HDPE and analyzing the properties that change at different blending ratios (Hossain et al., 2022; Kuan et al., 2022).

Comparison of HDPE with Other Recycled Poly-Olefins

The majority of plastic materials are recyclable and using them to create useful products is one way to lessen the environmental pollution. For this project, flat plates were created by hot pressing HDPE and LDPE into smaller pieces. To learn more about the material properties of both materials and whether they would be suitable for use in domestic products like bricks, pavement, and roof tiles, they were combined with industrial waste sawdust. Recycled HDPE was shown to have 10% worse impact resistance and tensile properties than virgin polymers, but better than recycled LDPE. Comparing samples with additional sawdust filler to those with just the original polymers, we found that the stiffness was higher, and the impact resistance was lower. In comparison to LDPE specimens. The increased density and modulus of the HDPE specimens were responsible for the higher wave velocity and lower attenuation that was seen during the ultrasonic experiment. The ultrasonic velocity, which represents the elastic modulus, and the attenuation, which represents the inhomogeneity of the microstructure, varied in opposition if the original specimens had more fillers. According to this study, recovered HDPE and LDPE exhibit mechanical qualities that are more similar to their virgin counterparts and offer a significant promise for plastic recycling (Nikiema and Asiedu, 2022; Zhu et al., 2022).

Due to their low cost, lightweight, and durability, HDPE and LDPE thermoplastic wastes were utilized. HDPE is extremely durable and does not easily degrade when exposed to sunlight, intense heat, or extreme cold. Although it is less harmful than other polymers and is typically used to make plastic bags, LDPE is not frequently recycled. The two plastics that are the simplest to recycle are these two. Even after being recycled into a new form, their special qualities allow them to be molded into a range of items for different uses. For the first time, research has demonstrated that the characteristics of plastic are unaffected by the molding and shredding processes during reuse. According to this result, this plastic may theoretically be recycled several times without degrading any of its original properties. This study aims to investigate the newly produced HDPE and LDPE from waste bottles in terms of their material properties and appropriateness for usage in domestic applications (Boesen et al., 2019; Gong et al., 2020).

Recycled HDPE specimens often outperform recycled LDPE in terms of tensile strength, impact resistance, and water resistance. It was discovered that the sawdust filler causes the elastic modulus of both forms of plastic waste to increase while decreasing impact resistance. Using the ultrasonic non-destructive method, increased wave velocity and reduced attenuation were achieved in HDPE specimens. The tensile and impact characteristics of specimens without extra filler can be ascertained using this ultrasonic technique. Better impact resistance is shown by a lower attenuation coefficient and a larger elastic modulus, respectively, with increasing velocity. The characterization of the recycled HDPE and LDPE specimens used in this study has provided significant insight into figuring out how recyclable they are and if they are suitable for use in domestic applications like bricks, pavements, and roof tiles. Therefore, by recycling plastic wastes, pollution of the air, water, and land may be reduced. This can help everyone have a more sustainable and environmentally friendly future (Bianchini and Rossi, 2021; Michaliszyn-Gabryś et al., 2022).

Due to a growth in both demand for and production of polymers, recycling of these materials is very popular nowadays. The thermomechanical processes used to recycle these materials change their general mechanical properties. In the current study, the ability of high-density polyethylene (HDPE), which is the most recycled material worldwide, can be recycled for use in additive manufacturing is the main focus. Through the use of a method that separates the thermomechanical treatment from other factors like age, and contamination. To determine the effect of ongoing recycling on the structural, mechanical, and thermal properties of HDPE polymer, a thorough investigation was carried out. Fused filament fabrication specimens created from both virgin and recycled materials underwent experimental testing and assessment. According to the results, HDPE is a good polymer for cyclical use since its overall mechanical behavior is generally enhanced through recycling stages for a set number of repetitions. As the number of extrusion cycles increased, the crystallinity of the HDPE polymer decreased, and throughout repeated extrusions, branching and cross-linking took precedence over chain scission, improving the mechanical properties. Using the experimental 3D printing settings utilized in this study, it was also shown that HDPE filament and recycled filament may be used in 3D printing applications, enhancing mechanical stability and toughness to 3D printed products without significant warping or other printing issues (Correa et al., 2022; Nikiema and Asiedu, 2022).

Life-Cycle Assessment of Recycling Postconsumer Polyethylene Terephthalate and High-Density Polyethylene

HDPE and PET mechanical recycling in Jordan aims to assess the combined environmental performance of both substances. The life-cycle assessment (LCA) method is used to investigate the potential environmental consequences of post-consumer HDPE and PET recycling. It assesses the total energy requirements, sources of energy, all types of pollutants, and waste generated between the post-consumer plastic recycling to produce PET and HDPE resin. System expansion and recycling cutoff allocations are employed as methods. The results of recycling "cut-off" and "system expansion" are compared to those obtained when the fresh resin is produced. A decrease in the consumption of non-renewable energy of between 40 and 85 percent and a reduction of emissions of greenhouse gases between 25 and 75 percent can be achieved, based on the framework to facilitate employment. Recycling PET and HDPE have considerable environmental benefits over virgin, single-use HDPE and PET, according to two allocation methodologies (Agovino et al., 2021; Bauer et al., 2022). LCA is a worthwhile method for assessing the potential benefits of recycling programs. It considers material and energy usage, pollutants in the environment, and waste disposal as it tracks each activity from raw materials extraction to the restoration of wastes to the earth. LCA has been utilized effectively for waste plastics recycling mechanically by contrasting several disposal choices. It was shown that mechanical waste recycling of plastics is superior to incineration and dumps as long as a certain recycled material replacement ratio is met. The thermal and mechanical characteristics of plastic trash have reportedly changed as a result of recycling. The conditions under which materials are processed and the caliber of the finished goods may change as a result. How to raise the caliber of recycled materials has been the subject of several studies. The grade of mechanically recycled HDPE and PET is determined by the waste stream's purity. Although recycled plastics hold great promise for usage in a variety of sectors, little is known about the environmental impact of creating recyclable materials or blends for rising applications or even everyday uses (Bianchini and Rossi, 2021; Zhu et al., 2022). LCA data illustrates that the total energy needs for pet flakes are 14 to 17 percent of virgin PET flakes and 57 percent of virgin resin, respectively. These results were obtained using the "cut-off" recycling technique and the "system expansion" approach. Recycled HDPE pellets for HDPE need 12 to 13% of the energy of virgin HDPE resin when recycled using the "cut-off" method, however, they need 62% when recycled to use the "system expansion" method. In addition, it is possible to conclude from the LCA results that recycling PET and HDPE can cut emissions of greenhouse gases by 25 to 75 percent. Compared to virgin materials made for single-use, recycled PET and HDPE provides significant environmental benefits and can boost eco-efficiency, according to LCA research (Nikiema and Asiedu, 2022; Williams and Rangel-Buitrago, 2022; Zhu et al., 2022).

Experiment

Collection

The collection of plastic waste material is a simple process. There are a lot of different ways of collecting plastic waste material. Locally, hundreds of scavengers do this job. Different collectors work the collection of all types of plastic from the surrounding. They can sell these waste plastic materials to different recyclers as their requirement. Our concern here is about the waste of HDPE plastic. These plastics contain other contaminants that disturb the smooth recycling process. The collected HDPE material needs to remove all the contaminant particles. Different batches of collected HDPE plastics contain different contaminants. Their compositions are also different.

Sorting

Sorting plastic is the biggest problem faced before recycling. It affects the properties of recycled products. Usually, the collected waste plastics may contain PVC, PP, PET, LDPE, and HDPE. The plastics usually contain 15%

of PVC which decreases the strength of HDPE. There is also a need to remove small impurities of PP, PET, and LDPE as they also affect the strength of the final product. The consideration here is to sort HDPE material from plastic waste material. For product sustainability, it is important to remove all the present impurities from the sorted material. The presence of impurities damages the production behavior of a material. Using this sorted material for manufacturing bottles of FMCG products, they must have a strength near that of the virgin HDPE.

Manual Sorting

Plastics must be manually sorted, and the operators must be able to visually identify each plastic by its distinctive size, color, form, and trademark. Whenever the substance to be separated is huge, this procedure is employed. Although this method is based on material identification codes, it cannot be disregarded since it has a high potential for human mistakes. This process is very much cheap compared to the other methods.

X-Rays and IR Sorting

Based on mixed plastic recycling research conducted by Waste Resource Action Program, NIR may effectively eliminate PLA biomaterials and carbon boards from a stream of mixed packaging (WRAP). In this process, near-infrared rays are used to irradiate the unsorted, unidentifiable plastic (600 to 2500 meters in wavelength). The speed of recognition for this is really quick. When exposed to near-infrared light waves, different polymers reflect an identifying spectrum, and this technique may successfully distinguish between various polymers. Whereas NIR reflectance spectroscopy offers several benefits for classifying and identifying plastic resins, it is not the best method for recognizing dark-colored plastics (Bauer et al., 2021; Shanker et al., 2022).

Air Sorting

The flakes of the comingled material, grounded to a ¼" to ½" size, are fed vertically into the air, affecting the lighter particle to be separated from heavier ones, and the material is sorted based on the specific gravity.

Mechanical Sorting

Although mechanical sorting has been successful in the mining sector, there hasn't been much advancement in the creation of mechanical sorting systems for recycling plastic. The mechanical sorting method makes use of centrifugal force, relative density, elasticity, particle shape, select shredding, and mechanical properties. For separating by centrifugal force, it will be necessary to design a bowl-type centrifuge with improved separating performance and less residual moisture content that divides polymers according to their specific gravity. In comparison to gravity-based separation, centrifugal sorting is much quicker and results in significantly drier plastic fractions. A cutting-edge method dubbed Hybrid-Jig, which is built on jigging and flotation has been developed in the area of mechanically sorting plastics. Jigging modifies the apparent specific gravity by adding air bubbles to the particle bed.

Crushing and Mixing

After the sorting and cleaning of the HDPE, we need to crush it into a crusher. The crusher will crush it into small tinny granules so that mixing of good quality can be achieved. As going to check the properties of 100% PCR, so leave some part of the r-HDPE for testing it and mix the remaining PCR with the V-HDPE in different percentages. Here, took three percentages which are 25% PCR and 75% V-HDPE, 30% PCR and 70% V-HDPE, and 50% PCR and 50% V-HDPE.

Extrusion

The process of heating nearly any plastic, whether it be in the form of powder, beads, flakes, pellets, or a mixture of these, is known as extrusion. The hopper of the extruder receives this plastic. The plastic is melted by the plasticity, a rotating spiral screw inside a heated barrel (cylinder), which is used by the extruder. To create the required continuous product form, the molten plastic is then pressed through a die. The extrusion process is depicted in Figure 1 in a much-condensed schematic. The feed is charged into the hopper after a manual mixing of different percentages of HDPE and r-HDPE where they are continuously mixed with the mixer whose speed can be controlled. The feed is charged into the barrel at a constant of 15rev. m^{-1}. The barrel speed is also fixed at 80 revs. m^{-1}. The temperatures of the barrel are set in the range of 165-200°C. The temperature on the die is 200°C. Different pressure and torques are created at the barrel depending on the mixture that is working.

Table 1 shows the overall readings of the extruder. After the extrusion, the material is passed through a crusher after passing through a water bath to cool the material and then crushed into tiny size granules. The speed of the crusher is fixed to 6m.min^{-1}.

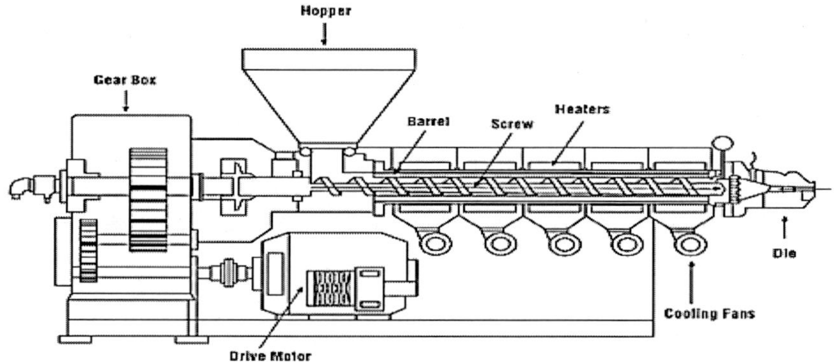

Figure 1. Extrusion machine.

Table 1. Readings for extrusion

Percentages	25% PCR and 75% V-HDPE	30% PCR and 70% V-HDPE	50% PCR and 50% V-HDPE
Barrel temperatures (°C)	165 – 200	165 – 200	165 – 200
Die temperature (°C)	200	200	200
Feed RPM (rev.m^{-1})	15	15	15
Barrel RPM (rev.m^{-1})	80	80	80
Barrel torque	25-27	21 - 25	17-21
Barrel pressure	52-55	52 - 54	51-54

Hot Press for Sample Manufacturing

After getting the pellets from extruders, the material is processed on a hot press machine for manufacturing of standard test sample that was used for mechanical characterization. The machine consists of an upper plate and a lower plate. These plates are used to provide heating and pressure to the sample placed in between them. The mold is made up-off metal which can bear a temperature of about more than 300°C. The mold consists of 3 to 4 cavities of dumbbell shape in which the sample is filled. Each cavity is of 3mm thickness, 13mm breadth, and 24mm length. Cover the sample containing the mold with metallic sheets covered with aluminum foil. The cavity must be filled with granules of material otherwise, there are many chances of voids in the sample. The plates are heated at a temperature of about 190°C. A pre-heat time of 15 minutes is provided to the machine to reach the temperature of 190°C. After reaching the temperature, place the mold between the plates. First, compress the plates to pre-heating the sample for 1 minute under low pressure. After pre-heating, release the pressure and apply a pressure of 100 bars, and then release. Repeat the same thing 3 times. Then press the plates under a pressure of 200 bars and leave them under that pressure for 15 minutes. Release the pressure after 15 minutes and remove the mold from the plates and leave it for cooling. It will take about 10 to 15 minutes to cool down. Then separate the sample from the mold and repeat the process for the other samples. Mold must be neat and clean and filled to get good quality of the sample. Check whether the sample has voids or not by passing light through it. Select only those samples which have no voids so that testing can be performed with good efficiency. The cycle time of the compression molding is about 40 – 45 minutes.

Testing

Testing is one of the important parts of any experiment. A small mistake generates the error. Each instrument must be carried very carefully. All safety gadgets must be worn while performing the testing.

Density Test

HPDE floats on the water, therefore, cannot find the density by using the technique of the volume, need to do it by the technique of mass of the sample. As the sample cannot be dipped in the solution, attach it with a small beat. First, take the mass of the sample in the air. Then take a small beaker and fill it with distilled water. Take the mass of the beat in water by dipping it into water. Then attach the sample with a beat and measure the mass of the beat and sample. Subtract the mass of the beat from the mass of the beat and sample to get the mass of the sample in water. Then by using the above formula, calculate the density of the sample. Repeat processes three to four-time with a different sample to get the density of the sample and take the average of the readings. Repeat the process for all the samples for the density of each.

Melt Flow Index Test

The melt flow index usually identified as MFI is one of the basic and important properties of polymers. The MFI is the amount of polymer flowing in grams per 10 minutes. The machine used for MFI is MFI ASTM D-1238. The apparatus consists of a barrel containing the granules of the polymer. There is a piston by which the melted polymer is forced to flow. A die of a thin hole is attached to the barrel. All the parts of the machine must be clean to get good quality results. First, clean the apparatus and put the initial requirements like the density, pre-heat, and time. Allow the barrel to reach the required temperature and then put the polymer into the barrel. Start the pre-heating of the polymer. It should be 300 seconds. The barrel must be in the range within the pre-heat time. If it is high in the required range or lowers the range, the testing will be aborted by the machine automatically. If the polymer present in the barrel is in the range, place the load of 2.16 kg onto the barrel. The melted polymer starts to flow from the die. The range should be less than 24mm to 25mm. The test will automatically stop when the polymer ends in the barrel. Note down the readings from the screen.

Tensile Test

Tensile testing measures the strength essential to break a plastic sample and the extent to which the sample expands or expands to the point of the fracture using ASTM D3039. First of all, start the system and set the gauge length (grip to grip length) to 50mm. The machine consists of gauges/grips between which the sample is to be attached. Calculate the thickness by using a screw gauge and the width by using the Vernier caliper. Note down the readings and put them into the system. Set the Stretching speed to 20 mm/min and push the start button. The machine will do the stretching and perform the test automatically. You must need to wear goggles and a helmet. The experiment will stop automatically when the limit of extension of the sample ends or the breaking of the sample occurs.

Flexural Test

The flexural test was performed on a Universal testing machine made by Zwick Roell Germany, as per the standards of ASTM C293/C293M-16. Flexural testing measures the force required to bend a plastic beam to determine how stiff or resistant it is to bend material. Flex modulus describes the material's capacity to bend before irreversibly deforming. For snap-fit assemblies or plastic lock arms, the arm must bend to facilitate appropriate connection seating and then flex back into position to secure the connection in place. If the locking mechanism is composed of fragile material, it will be more likely to shatter when it is bent. Flex testing for support beams, for instance, reveals how much weight the beams can sustain before bending. As a result, a rigid or stiffer material is more suited for such an application.

Izod Test

The Izod test measures the energy sustained by a substance under a high-speed impact at a high strain rate. In the Izod test, the specimen is placed in a cantilever manner. The test can be performed either with the notch or without the notch. A V-Shaped notch is preferred for the Izod test using ASTM D256. After creating the notch from notch making machine the sample is attached with a vice in a cantilever manner. A V-shaped notch, facing forward, may be seen in this specimen. After that, the hammer of different energy levels is raised to a certain height and released. The hammer hit the sample from the top and broke the sample. The energy by which the sample was broken will be observed. When divided by the area of the sample the impact strength can be calculated. The impact strength depends on the length of the arm, and the height from which it is dropped. The arm's length and the height at which it is dropped both affect how quickly the impact occurs. The load of the hammer/pendulum used in the test is 0.95kg. The specimen used was of a rectangular shape of dimensions 45.35mm x 2.8mm x 0.3mm. The velocity of the hammer when it hits the sample is 3.46 m/s.

Discussion

Density

The difference in the densities of recycled and virgin HDPE is very small. The virgin HDPE has a density of 0.9151gm/cm^3 and the PCR has a density of 0.9453 gm/cm^3. When increases the virgin content in the PCR, the density tends to decrease. Figure 2 is representing the trend of the densities for 100% recycled material, and virgin material and the trend of densities for different compositions of recycled and virgin material. According to the trends it has been observed that the PCR is showing greater density as compared to virgin material. This may be due to impurities that might be present in PCR material. With adding virgin material into PCR the density of PCR shows a decreasing trend. Overall there isn't any major difference in densities for all compositions that restricts its use in cosmetic bottles (Kuan et al., 2022; Morgan et al., 2022).

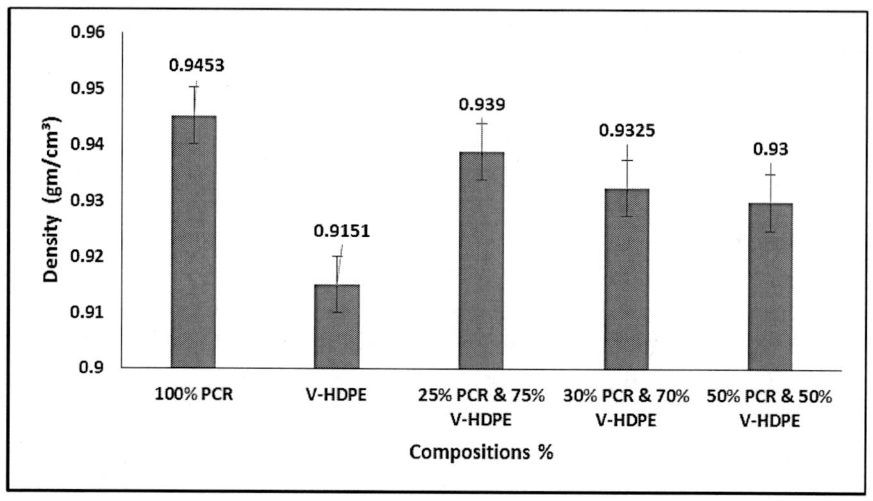

Figure 2. Effect of density on different compositions of PCR and V-HDPE.

Melt Flow Index

The melt flow index increases for the PCR as compared to virgin material. There might be a different reason for increasing the mass flow rate of PCR material. When the material is processed and used in the designated application, some impurities may be trapped in the product. These impurities may change the flow behavior of the polymer and increase the melt flow rate of PCR material. As increases the recycling content in the V-HDPE, its MFI increases a little bit, which may be negligible (Gall et al., 2020; Sitadewi et al., 2021). Figure 3 clearly shows a small change of 0.003 gm/cm^3 from 25% PCR to 50% PCR in V-HDPE. The flow rate can be controlled by adjusting shear stress and temperature.

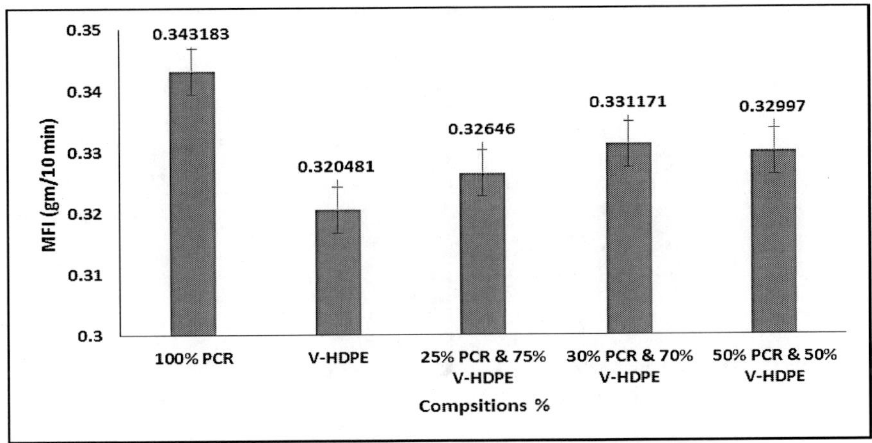

Figure 3. Effect of MFI on different compositions of PCR and V-HDPE.

Tensile Strength

Figure 4 compares the tensile strength of the v-HDPE and r-HDPE. There is a slight difference in the stress applied to the v-HDPE and r-HDPE. R-HDPE is more ductile and hence will break without elongating as compared to the v-HDPE which elongates 120%. The tensile test is only performed on PCR material and virgin material to check the effect of tensile forces. The rest of the compositions are not tested in detail as the main purpose of setting different compositions is to use the material for the manufacturing of cosmetic bottles. Tensile properties are not so crucial for these bottles. Three different samples of PCR and virgin material are tested. According to the obtained results it has been observed that upon applying tensile stress the PCR material shows low resistance and its strength decreases. The decrease in strength may result from short chains that provide lower resistance to the application of tensile stress (Bauer et al., 2021; Cecon et al., 2022).

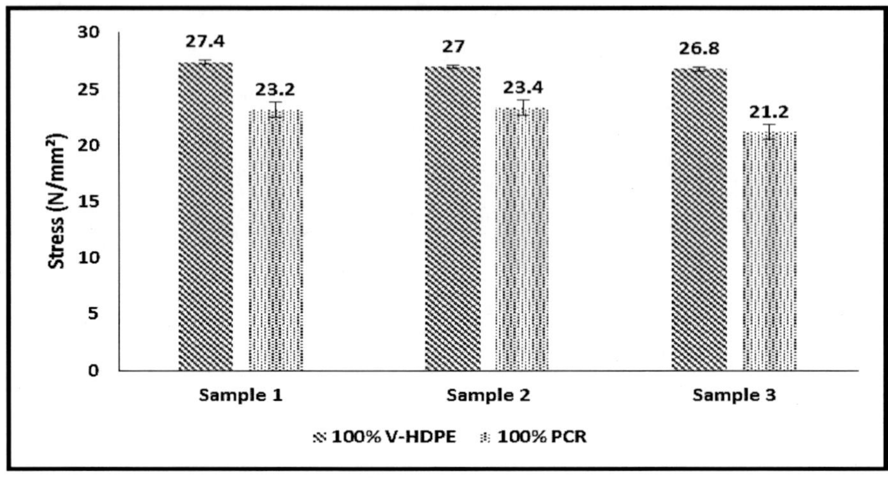

Figure 4. Effect of stress on a different sample of PCR and V-HDPE.

Flexural Strength

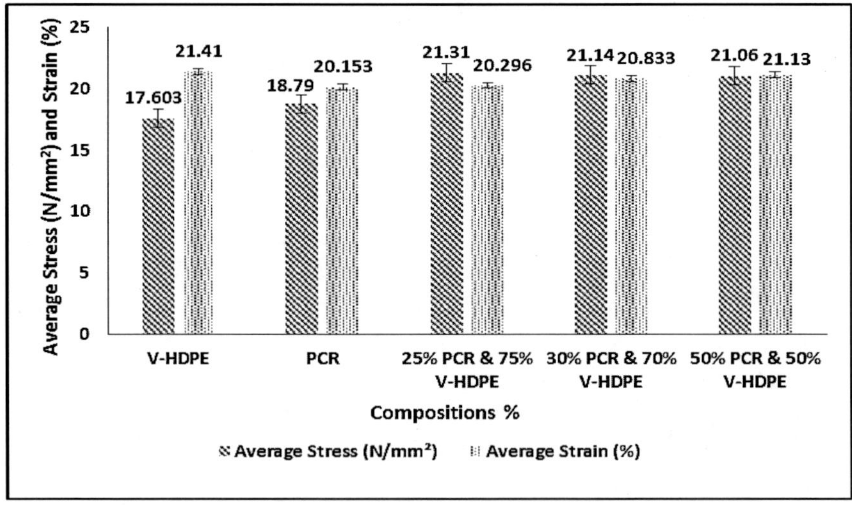

Figure 5. Effect of average stress and strain on different compositions of PCR and V-HDPE.

For all the flexural test samples the strain rate was 50mm/min. Figure 5 shows the results of stress and strain of each composition when they are tested upon the application of bending force. Flexural property is important for bottles, it tells us the maximum bending stress. Because of the stiffness, recycled material shows better flexural properties than virgin. When adding that recycled content in the v-HDPE, its flexural properties deviate a small. According to the results it has been observed that the virgin material resists the stress of 17 N/mm^2 while the composition of 25% and 75% resist the stress of 21 N/mm^2. So this composition might work for the manufacturing of bottles (Cecon et al., 2022; Garofalo et al., 2019; Morgan et al., 2022).

Izod Impact Strength

In Figure 6, the impact strength of all the compositions is shown. Impact strength decreases to increase the recycled content due to an increase in the ductility of the material. The recycled content has a weak chain. The test was performed with un-notched samples. When the hammer hits the sample, all the stress act on weak areas, and samples was usually broken from random areas. However, the composition of 25% and 75% is closer to the value of virgin material (Cecon et al., 2022; Juan et al., 2020; Sitadewi et al., 2021).

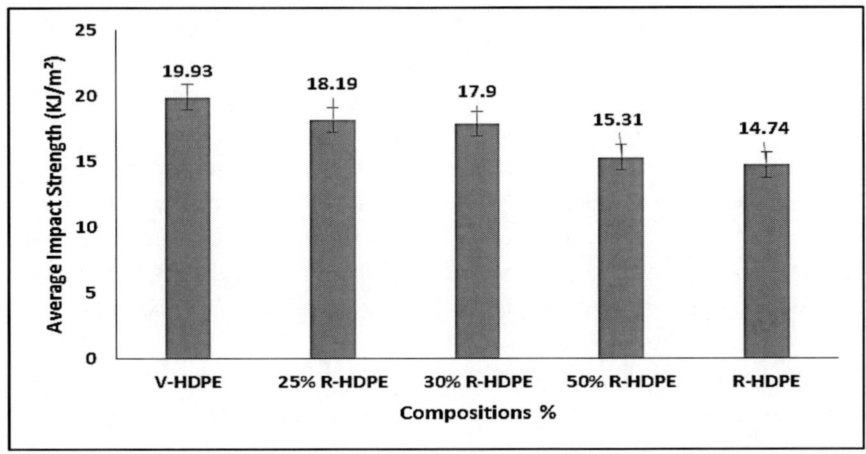

Figure 6. Effect of impact strength on different compositions of PCR and V-HDPE.

Conclusion

Recycling plastic is one of the most diversified areas and has extremely important environmental concerns. The use of recycling material in a good application will also solve environmental problems. The core purpose of this project is to recycle HDPE which is one of the most used plastic resins in packaging. The starting point of the process is to sort out the collected plastic waste which contains materials other than HDPE like PP, LDPE, PS, and PVC which are then sorted using a Near-infrared sorter, washed and cleaned, and after it is recycled in a reactor having an agitator. After segregation and separating of HDPE from all the waste, the material is put into crushers so that it can be converted into small segments. After crushing different formulation has been set for waste and virgin material. All these formulations were put in a twin-screw extruder for better melt mixing and pelletization. The co-rotating intermeshing twin screw extruder is used for melt mixing and pelletization. The parameters like screw speed, the temperature of different zones, pressure at the die head, and torque have been carefully monitored to check the effect of PCR material on processing. After the complete pelletization, the material was processed in a hot press and a sample for mechanical testing was manufactured. The dimensions of the mold and specimen follow international standards such as ASTM and ISO. Promising results have been obtained when the material is tested concerning its flow behavior and its mechanical properties. A melt flow index test was performed for the measurement of the melt flow rate and tensile, flexural, and impact test was performed to measure the mechanical properties. According to the overall results, the MFI and flexural properties increases by adding PCR material while tensile and impact properties decrease but the formulation of 25% and 75% shows results that are very close to virgin material in all testing condition. So, the result of the finding was successfully applied by Unilever Pakistan in their shampoo bottle with 25% recycled content. The next challenge is to apply this in their white shampoo bottles the resin that gets after recycling is not 100% white so the more optimized the process of sorting is the more suitable it will be for practical applications.

This study developed an environmentally and economically efficient method of recycling HDPE that involves collecting, sorting, cleaning, crushing, extrusion, sampling, and testing. After conducting numerous studies with various compositions of virgin HDPE and recycled HDPE, discovered that the composition's physical properties are quite similar to those of the virgin HDPE. The experiment focused on three compositions: 25% PCR and

75% V-HDPE, 30% PCR and 70% V-HDPE, and 50% PCR and 50% V-HDPE. There can be a slight variation in the color of the bottles; however, this can be minimized by using more master batches (whitener). Moreover, the collection and sorting procedure has a substantial impact on the quality of the recycled plastic. Therefore, to obtain the best quality and most varied composition of recycled HDPE in their products, it is necessary to concentrate on and maximize the collection and sorting process.

References

Agovino M, Ferraro AMusella G. Does national environmental regulation promote convergence in separate waste collection? Evidence from Italy. *J Cleaner Prod* (2021) 291(1):1-17.

Bauer, AS, Leppik, K, Galić, K, Anestopoulos, I, Panayiotidis, MI, Agriopoulou, S, Milousi, M, Uysal-Unalan, I, Varzakas, T, & Krauter, V. Cereal and Confectionary Packaging: Background, Application and Shelf-Life Extension. *Foods* (2022) 11(5):1-28.

Bauer A S, Tacker M, Uysal-Unalan I, Cruz, RMS, Varzakas, T, & Krauter, V. Recyclability and Redesign Challenges in Multilayer Flexible Food Packaging-A Review. *Foods* (2021) 10(11):1-17.

Bianchini ARossi J. Design, implementation and assessment of a more sustainable model to manage plastic waste at sport events. *J Cleaner Prod* (2021) 281(1):1-26.

Boesen S, Bey NNiero M. Environmental sustainability of liquid food packaging: Is there a gap between Danish consumers' perception and learnings from life cycle assessment? *J Cleaner Prod* (2019) 210(1):1193-1206.

Cecon V S, Curtzwiler G WVorst K L. Influence of Biofillers on the Properties of Regrind Crystalline Poly(ethylene terephthalate) (CPET). *Polymers* (2022) 14(15):1-15.

Colwill J, Simeone A, Gould O, Woolley, E, & Mulvenna, C. Energy-efficient Systems for the Sensing and Separation of Mixed Polymers. *Procedia* (2017) 62(1):512-517.

Correa C A, De Oliveira M A, Jacinto C, & Mondelli, G. Challenges to reducing post-consumer plastic rejects from the MSW selective collection at two MRFs in Sao Paulo city, Brazil. *J Mater Cycles Waste Manag* (2022) 24(3):1140-1155.

Gall M, Wiener M, Chagas de Oliveira C, Lang, RW, & Hansen, EG. Building a circular plastics economy with informal waste pickers: Recyclate quality, business model, and societal impacts. *Resour Conserv Recycl* (2020) 156(1):1-11.

Garofalo E, Di Maio L, Scarfato P, Di Gregorio, F, & Incarnato, L. Nanotechnology-Based Strategy to Upgrade the Performances of Plastic Flexible Film Waste. *Polymers* (2019) 11(5):1-17.

Gong Y, Putnam E, You W, & Zhao, C. Investigation into circular economy of plastics: The case of the UK fast moving consumer goods industry. *J Cleaner Prod* (2020) 244(1):1-13.

Hossain R, Islam M T, Shanker R, Khan, D, Locock, KES, Ghose, A, Schandl, H, Dhodapkar, R, & Sahajwalla, V. Plastic Waste Management in India: Challenges, Opportunities, and Roadmap for Circular Economy. *Sustainability* (2022) 14(8):1-34.

Juan R, Domínguez C, Robledo N, Paredes, B, & García-Muñoz, RA. Incorporation of recycled high-density polyethylene to polyethylene pipe grade resins to increase close-loop recycling and Underpin the circular economy. *J Cleaner Prod* (2020) 276(1):1-12.

Kuan S H, Low F SChieng S. Towards regional cooperation on sustainable plastic recycling: comparative analysis of plastic waste recycling policies and legislations in Japan and Malaysia. *Clean Technol Environ Policy* (2022) 24(3):761-777.

Michaliszyn-Gabryś B, Krupanek J, Kalisz M, Smith, J. Challenges for Sustainability in Packaging of Fresh Vegetables in Organic Farming. *Sustainability* (2022) 14(9):1-29.

Morgan D R, Styles DThomas Lane E. Packaging choice and coordinated distribution logistics to reduce the environmental footprint of small-scale beer value chains. *J Environ Manage* (2022) 307(1):1-11.

Nikiema JAsiedu Z. A review of the cost and effectiveness of solutions to address plastic pollution. *Environ Sci Pollut Res Int* (2022) 29(17):24547-24573.

Rosenboom J G, Langer RTraverso G. Bioplastics for a circular economy. *Nat Rev Mater* (2022) 7(2):117-137.

Shanker R, Khan D, Hossain R, Islam, Md. T, Locock, K, Ghose, A, Sahajwalla, V, Schandl, H, & Dhodapkar, R. Plastic waste recycling: existing Indian scenario and future opportunities. *Int J Environ Sci Technol (Tehran)* (2022) 19(4):1-18.

Siracusa VBlanco I. Bio-Polyethylene (Bio-PE), Bio-Polypropylene (Bio-PP) and Bio-Poly(ethylene terephthalate) (Bio-PET): Recent Developments in Bio-Based Polymers Analogous to Petroleum-Derived Ones for Packaging and Engineering Applications. *Polymers* (2020) 12(8):1-17.

Sitadewi D, Yudoko GOkdinawati L. Bibliographic mapping of post-consumer plastic waste based on hierarchical circular principles across the system perspective. *Heliyon* (2021) 7(6):1-23.

Williams A TRangel-Buitrago N. The past, present, and future of plastic pollution. *Mar Pollut Bull* (2022) 176(1):1-20.

Zhu Z, Liu W, Ye S, & Batista, L. Packaging design for the circular economy: A systematic review. *Sustain Prod Consum* (2022) 32(1):817-832.

Chapter 2

Injection Molding and 3D Printers: Advances in Research and Applications

İdris Karagöz*
and Neşe Çakır Yiğit
Department of Polymer Materials Engineering, Yalova University, Yalova, Türkey

Abstract

Injection molding and 3D printing have become important technologies in the manufacturing industry in recent years. Both technologies offer a fast, precise, and cost-effective method in the production process. The difference between these two technologies comes from the methods of shaping materials. Injection molding is a process that produces high-quality, high-volume parts by injecting thermoplastic or thermoset materials into a mold under high pressure. On the other hand, 3D printers create a part by depositing materials layer by layer. This method allows for the production of customized parts using a wide range of materials. Both injection molding and 3D printing offer significant advantages in the manufacturing process. Injection molding is ideal for high-quality, high-volume production, and parts produced by this method offer advantages such as high durability, excellent surface quality, and precise dimensions. 3D printing is advantageous in customized production, allowing for the production of parts in different shapes and sizes. Additionally, 3D printing is an important tool in prototyping. Significant progress has been made in both technologies with new materials, new design features, and faster production methods being developed. These advances have increased the applicability of injection molding and 3D printing. As a result, both technologies offer different advantages for

* Corresponding Author's Email: idris.karagoz@yalova.edu.tr.

In: Injection Molding
Editor: Jose D. Phillips
ISBN: 979-8-89113-429-4

different applications, and their importance in the production process is increasingly recognized and expected to continue to develop in the future. This study aims to compare the advantages and disadvantages of injection molding and 3D printing and convey the latest developments in these methods.

Keywords: injection molding, 3D printers, manufacturing industry, material shaping, high-quality production

Introduction

Injection molding and 3D printers are two distinct technologies that play a significant role in transforming design into tangible products in the modern manufacturing industry (Karagöz et al., 2021). Both technologies enable the production of parts from a designed model. Injection molding involves the use of a mold, while 3D printers allow for production without the need for a mold. These methods provide opportunities for part production, but they employ different processes and serve different purposes.

Figure 1a presents a schematic representation of the steps involved in injection molding, which is commonly used for mass production. In this method, a molten polymer is injected into an injection mold designed with the desired shape of the object. The material cools and solidifies within the mold, and the final part is ejected from the mold using ejectors. Injection molding is known for its efficiency, high precision, the ability to produce multiple parts simultaneously, and the option to use different colors. Therefore, it finds wide application in the serial production of plastic components, toys, automotive parts, and various industrial products (Karagöz, 2021; Karagöz and Tuna, 2021). On the other hand, 3D printers utilize a layer-by-layer approach to produce objects, as shown in Figure 1b. They operate based on a digital model created using computer-aided design (CAD) software. The model is divided into a series of layers, which are then transformed into physical layers by the 3D printer controlled by software (Karagöz et al., 2021; Çakır Yiğit and Karagöz, 2023). Each layer is built by depositing or solidifying a specific material, such as plastic, metal, or ceramic. This process allows for the creation of objects with intricate geometries and internal structures. 3D printers are used for applications such as prototyping, pre-production testing, personalized manufacturing, and specialized applications. Figure 1b specifically illustrates the production of parts using polymer filaments in 3D printers.

Injection molding and 3D printers are important production technologies that enhance innovation and productivity in the manufacturing industry. Each technology has its own advantages and areas of application. Injection molding is a fast and efficient option for large-scale production, offering high precision, excellent surface finish, strength, and durability (Karagöz, 2018). It is widely used in the automotive sector, electronics, toys, packaging, consumer products, and many other industries (Kandemir et al., 2023). On the other hand, 3D printers excel in smaller-scale production, prototyping, and customized manufacturing, providing designers and engineers with greater design flexibility and creativity. They can easily produce features that are challenging with traditional manufacturing methods, such as complex geometries, organic shapes, and internal cavities.

These two methods can be used separately or together as complementary technologies at different stages of the production process. Combining injection molding and 3D printers can result in a more robust production process. For example, a prototype can be quickly produced using a 3D printer, undergo necessary quality checks, and once the part is approved, an injection mold can be created for mass production.

As technology advances, the boundaries between injection molding and 3D printing are becoming blurred. Advanced 3D printers are now capable of directly producing parts using materials similar to those used in injection molding. This convergence of technologies can accelerate the transition from prototyping to mass production while reducing costs. In conclusion, injection molding and 3D printers are fundamental technologies in part production within the manufacturing industry. As mentioned above, each technology offers different advantages and areas of application. The combination of these two technologies contributes to more efficient, flexible, and innovative manufacturing approaches, shaping the future of production methods.

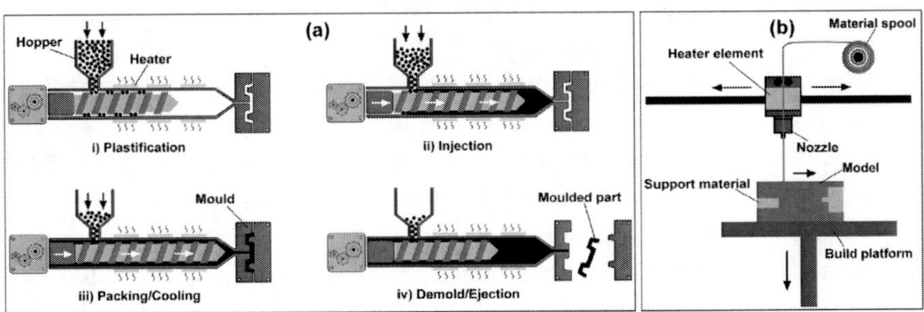

Figure 1. Schematic representation of part production with injection and 3D printers, (a) Part production with injection molding, (b) Part production with 3D printer.

This study aims to examine the recent developments in injection molding and 3D printing technologies, compare their advantages and disadvantages, and provide a resource for professionals working in this field. Additionally, it aims to contribute to the more effective and efficient utilization of technologies used in the conversion of design to product in the manufacturing industry. Firstly, the focus will be on the molding of parts using the injection molding method with thermoset, thermoplastic, and elastomeric materials. The properties of these materials, their applications in the injection molding process, and their advantages will be discussed. Next, the production of parts using thermoplastic filaments with 3D printers will be examined. The working principle of 3D printers and the layer-by-layer assembly of models designed using CAD software will be explained. Topics such as the advantages of this method, its use in the prototyping process, its role in personalized production, and the ease of creating complex geometries will be addressed. Subsequently, a comparison will be made between injection molding and 3D printing methods. The advantages and disadvantages of both technologies will be compared, and their different application areas will be emphasized. The suitability of injection molding for mass production features such as high precision, excellent surface finish, and the ability to mold multiple parts will be discussed. On the other hand, the advantages of 3D printers such as prototyping, customized production, and design uniqueness will be highlighted. In conclusion, this study aims to provide a resource for individuals interested in and working in the field by examining the recent developments in injection molding and 3D printing technologies, comparing the advantages and disadvantages of both methods and shedding light on their different applications.

Injection Molding of Thermoplastic Polymers

Thermoplastics are polymers that can be melted or softened repeatedly by heating, then solidified by cooling as they undergo physical changes. Injection molding is the most common way to manufacture parts from thermoplastic polymers. It is particularly suitable for thermoplastic polymers, which can be melted and solidified repeatedly without undergoing significant degradation. Injection molding offers several advantages, including high production rates, design flexibility, and cost-effectiveness (Xu et al., 2015; Wang et al., 2020; Zhao et al., 2021). Thermoplastic materials are a comprehensive class (see

Table 1) that provides a suitable polymer structure for almost any application (Heckele and Schomburg, 2003).

Table 1. A list of thermoplastic polymers commonly used in injection molding

Acronym	Full Name	Temperature stability (°C)	Properties	Structure	Usage area
ABS	Acrylonitrile butadiene styrene	100°C	Opaque	Amorphous	Drainage systems, plastic clarinets, golf club heads, car accessories, household gadgets, and LEGO blocks
PE	Polyethylene	100-150°C	Transparent, milky-opaque, or opaque	Variable crystalline structure	Plastic bags, films, storage units (including bottles), and plastic membranes
PC	Polycarbonate	130°C	Highly transparent	Amorphous	Greenhouses, compact discs, optical lenses, medical instruments, automobile accessories, and smartphones
PA	Polyamide	80-120°C	Good electrical properties, durability, abrasion resistance, and chemical resistance	Semi-crystalline	Healthcare products, automotive parts, sports equipment, clothing, shoes, and industrial elements
PP	Polypropylene	110°C	High melting point, fracture resistance, and excellent impact strength	Semi-crystalline	Kitchenware, sportswear, floor coverings, and automotive electrical systems
HIPS	High impact polystyrene	170-235°C	Affordable, good dimensional stability, and stiffness	Translucent	Containers for various food products

The general steps of the injection molding process are as follows: The mold cavity is closed, evacuated, and heated above the polymer glass transition temperature by an injection unit and the viscous polymer is pressed into the mold, and the polymer (and tool) is cooled and the polymer removed from the mold (Heckele and Schomburg, 2003; Wilczyński et al., 2022).

The first step in injection molding is designing the mold. The mold consists of two halves, a stationary half (cavity) attached to the fixed side of the press and a moving half (core) attached to the moving clamp side of the press. They fit together to form a hollow cavity in the desired shape of the final product. To begin with, thermoplastic pellets or granules are fed into a hopper, which feeds them into a heated barrel. Inside the barrel, the pellets are melted by heat and by the mechanical action of a screw. Once the plastic is melted and homogenized, the injection molding machine injects the molten material into the mold under high pressure. The molten plastic fills the cavity and takes the shape of the mold. The pressure is maintained until the plastic solidifies. The plastic cools and solidifies inside the mold. Cooling can be accelerated by circulating water or other cooling fluids through channels in the mold. The cooling time depends on the thickness and geometry of the part. The mold is opened once the plastic has solidified and the part is ejected. Ejection can be done using ejector pins, compressed air, or other mechanical means. The cycle repeats for the next part (Bendada et al., 2004; Fu et al., 2020).

Injection molding provides numerous benefits for thermoplastic polymers:

1. Injection molding is an automated process that efficiently produces substantial quantities of parts with high efficiency and high production rates, making it ideal for mass production.
2. It offers flexible design options, allowing complex and intricate part designs that include thin walls, intricate details, and various shapes. It also facilitates the integration of multiple parts or features into a single component.
3. It offers a versatile range of materials, supporting a wide variety of thermoplastic polymers, including commonly used ones such as polyethylene (PE), polypropylene (PP), polystyrene (PS), acrylonitrile butadiene styrene (ABS), specialty engineering plastics such as polycarbonate (PC), polyamide (PA) and polyether ether ketone (PEEK).

4. It is cost-effective for large-scale production runs due to the high speed and repeatable nature of the process (Fu et al., 2020; Jeong et al., 2022).

Nevertheless, it is essential to bear in mind certain factors:

1. The creation of molds can be expensive, particularly for intricate designs. However, these costs are typically spread out over the production volume.
2. It imposes specific design considerations, including draft angles, maintaining consistent wall thickness, and accounting for material flow and shrinkage.
3. Different thermoplastic polymers exhibit distinct properties and behaviors during molding. Selecting the appropriate material should be based on the specific requirements of the part, including mechanical strength, temperature resistance, chemical resistance, and aesthetic considerations (Fu et al., 2020).

Injection molding involves a range of process parameters, such as melt temperature, injection pressure, and cooling time. Incorrectly configuring these parameters can result in excessive shrinkage, a prevalent problem in thermoplastic polymer injection molding. Shrinkage arises due to the inherent material contraction during the cooling phase. A significantly high pressure is maintained within the molding cavity throughout the cooling process to ensure proper mold filling and counteract the volume reduction caused by thermal shrinkage and potential solidification shrinkage in semi-crystalline polymers. In recent years, extensive research has been carried out to optimize the molding process parameters to reduce the shrinkage and occurrence of warping in injection-molded products (Massé et al., 2004; Santis et al., 2010; Sun et al., 2019). Stresses resulting from warpage significantly impact the product's mechanical properties and are closely tied to the molding process parameters. Consequently, considering the influence of the molding process is crucial when describing the mechanical behavior of the product (Xu et al., 2015; Farotti and Natalini, 2018). Furthermore, the mechanical properties of the bulk material are significantly influenced by crystallinity, as crystals tend to be stiffer than amorphous materials. Additionally, orientation introduces anisotropy and other alterations in mechanical properties (Pantani et al., 2005).

Giboz et al. investigated the microinjection molding of thermoplastic polymers and compared their morphological differences with conventional injection molding. The results showed that the specific conditions used in microinjection molding led to smaller and more oriented crystal formations with a higher effect of flow-induced crystallization (Giboz et al., 2009). Liou and Chen conducted a study to examine how injection molding conditions affect the moldability of micro- and sub-micron structures using common engineering plastics like PMMA, PP, and HDPE. They also investigated these polymers' unique filling behaviors and molding characteristics in microinjection molding. The findings revealed that the molding characteristics in micro-injection molding are primarily influenced by the mold temperature, main injection pressure, and the specific polymer being used. Moreover, the mold temperature required for micro-injection molding significantly surpasses traditional injection molding (Liou and Chen, 2006).

Injection Molding of Thermoset Polymers

Figure 2 schematically illustrates the general structure of the screw barrel system and the molding process used in the injection molding of thermoplastics and thermosets. Thermoset materials such as urea-formaldehyde, phenol-formaldehyde, and unsaturated polyesters are commonly used in the injection molding of thermosets. Although there are differences, the injection molding of thermosets is similar to the injection molding of thermoplastics. Thermosets are materials that harden through a polymerization process. The injection molding of these materials is accomplished by injecting pre-prepared thermoset resins into a mold cavity using a screw (injection screw) with the aid of temperature and pressure. The first step in the injection molding of thermosets is the injection of the liquefied thermoset material into the mold cavity. In this process, temperature and pressure settings are adjusted in the machine for the shaping of thermoset materials. Once the desired temperature values are reached, the mold is closed, and the liquid thermoset material is injected into the mold cavity under high pressure with the help of the injection screw, taking the shape of the mold cavity. After the thermoset material hardens in the mold cavity, the mold is opened, and the part is removed from the mold. If applicable, secondary operations (such as cutting, drilling, etc.) are performed on the part to complete the production.

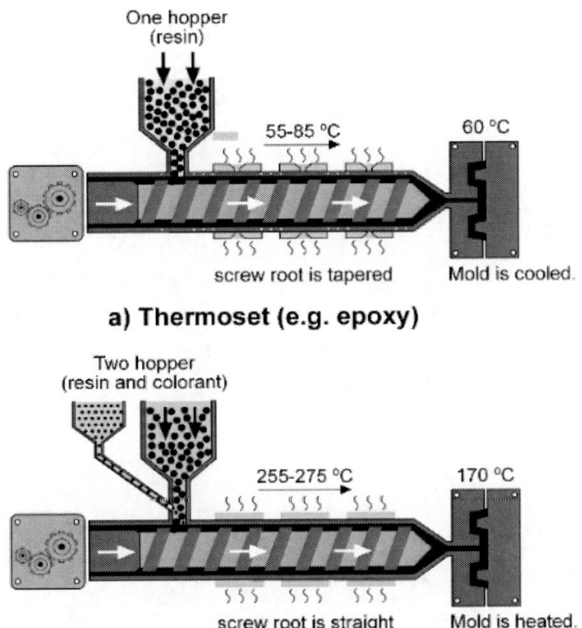

Figure 2. Schematic representation of injection molding of thermoplastics and thermosets.

The changing structure and bond formation of thermoset polymers with heat is schematically shown in Figure 3. The fundamental differences between the injection molding of thermoplastics and thermoset materials can be listed as follows:

1. The barrel temperature used in shaping thermoset materials is lower than the temperatures used in shaping thermoplastics. The main reason for this is that the liquefied thermoset resin needs a lower temperature to minimize the formation of cross-links during mold filling and to polymerize and harden within the mold.
2. After being filled into the mold cavity, thermoplastic polymers require cooling to maintain their shape, while in thermoset materials, the resin is kept heated within the mold to accelerate cross-linking.
3. The designs of injection screws are different for both materials. In thermoplastics, the screw thread heights generally increase towards

the tip of the screw. In thermosets, injection screws with uniform thread heights are used.

4. Thermoplastic materials can be produced in different colors by using different coloring agents during injection molding. However, coloring agents are not commonly used in thermosets.
5. Thermoset materials do not lose the geometry they acquire after molding, so it is not a problem to remove them from the mold while they are still hot. Thermoplastic materials, on the other hand, need to be cooled within the mold to maintain their shape before being removed from the mold.
6. Thermoset materials show less chemical interaction with the usage environment compared to thermoplastics.

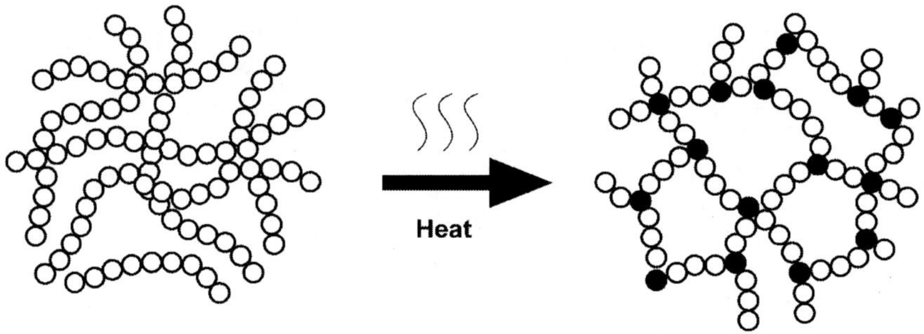

Figure 3. Insoluble bond formation with heat in a thermoset plastic.

The injection molding of thermoset materials is ideal for the production of parts that require high-temperature resistance, high strength, and chemical resistance. Therefore, injection molded parts made of thermoset materials are used in many industries such as automotive, aerospace, electronics, medical devices, and industrial equipment. However, the injection molding of thermoset materials is more challenging and costly than the injection molding of thermoplastics due to the complexity of the process. Opening the mold and removing the part take longer compared to thermoplastics, increasing production time. Additionally, the recycling of thermoset materials is more difficult, which can pose a significant waste management issue. The temperature values for the injection molding process of some thermoset materials are provided in Table 2.

Table 2. The injection molding process temperatures for some thermoset materials

Material	Nozzle	Cylinder	Mold	Material
Phenolic	80-100°C	55-75°C	170-190°C	110-130°C
Urea	85-115°C	65-80°C	130-150°C	120-140°C
Melamine	90-110°C	65-80°C	160-170°C	115-135°C
Unsaturated polyesters	70-100°C	50-70°C	170-200°C	70-100°C

Injection Molding of Elastomers

Elastomers have a behavior that lies between thermoplastics and thermosets, and they exhibit resilience to heat and solvents. Their ability to undergo significant shape changes, much like rubber, and exhibit considerable elongation with small stresses are their most important characteristics. Under external loads, they can undergo substantial elastic deformation (combining viscosity and elasticity in their structure), but they return to their original dimensions when the load is removed (Figure 4). They have a linear molecular structure and are linked together at specific points through cross-linking in the vulcanization process. There are various types of elastomers (Saçak, 2014). Some require vulcanization and are generally thermosetting. However, there are also thermoplastic types, such as ethylene-propylene diene monomer (EPDM), ethylene-propylene terpolymer (EPT), styrene-butadiene-styrene triblock copolymer (SBS), and thermoplastic polyurethane (TPU).

Their elastic properties make elastomers stand out as an ideal material for various applications. Injection molding is a highly efficient method used in the mass production of elastomers. Products manufactured using this method offer high precision, repeatability, and find applications in diverse industries such as industrial, automotive, medical, and electronics. Unlike thermosets and thermoplastics, the injection molding of rubber requires specialized heating and cooling equipment in the injection machines. While thermoplastics are fed in granule form and thermosets in liquid form, elastomers are fed in strips to the injection machine. The softened elastomer material moves along the injection barrel and is pressed into the mold with the help of an injection screw, where cross-linking reactions take place to complete the process (Akkurt, 2011). As the material is injected into the mold in a hot state, the vulcanization time is short. Through vulcanization in the mold, the mechanical properties of the parts are improved, resulting in non-sticky components that do not soften excessively when heated and exhibit high elasticity over a wide temperature range. Since vulcanization takes place during the molding process, the leftover sprues and production waste (flash, etc.) cannot be reused after molding.

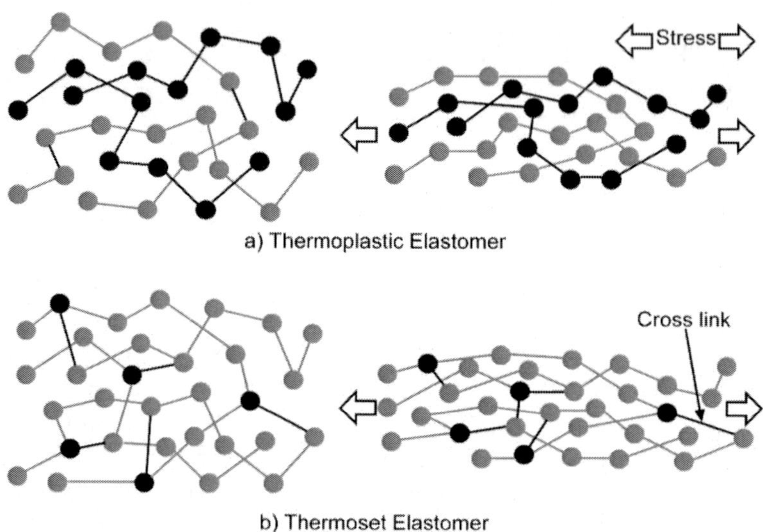

Figure 4. Thermoplastic and thermoset elastomers.

The structure of thermoset polymers, elastomers, thermoplastic polymers, and thermoplastic elastomers is schematically shown in Figure 5. The advantages of injection-molded elastomers include high flexibility, durability, abrasion resistance, and chemical resistance. However, they also have some disadvantages. For instance, elastomers can be prone to degradation under high temperature or high-pressure conditions, and they may be more sensitive to environmental factors.

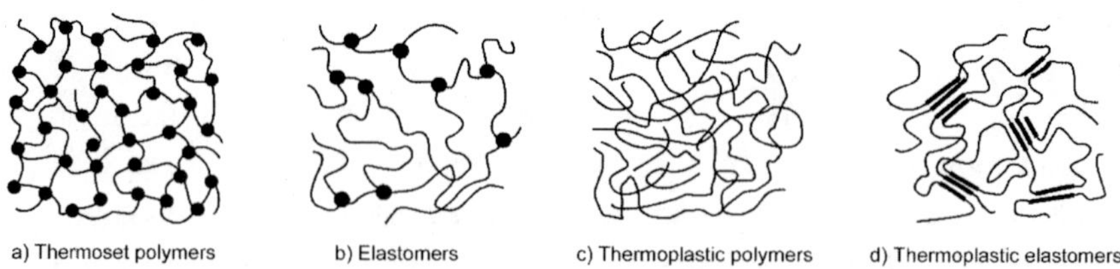

Figure 5. The structure of thermoset, elastomers, thermoplastic and thermoplastic elastomers.

3D Printing of Thermoplastic Polymers

Additive manufacturing (AM) or 3D printing technique is a process of creating three-dimensional objects by depositing material layer-by-layer basis. 3D printing allows for the creation of intricate and complex designs that may not be feasible with conventional manufacturing techniques (drilling, milling, extrusion, injection, and so forth). In the process, a thermoplastic filament is fed into a heated nozzle, which melts the material. The molten polymer is then extruded onto a build platform, layer by layer, following the desired design. The layer-by-layer approach in 3D printing provides meticulous control over the object's geometry and characteristics. Each layer is carefully deposited and solidified before adding the next layer, ensuring precise accuracy and maintaining the object's structural integrity (Blok et al., 2018; Karakurt and Lin, 2020; Schönhoff et al., 2021).

The basic logic of 3D printer technology is to turn objects designed in three dimensions into concrete objects by stacking thin layers on top of each other on a two-dimensional plane using CAD/CAM (Computer Aided Design/Computer Aided Manufacturing) programs. Parts designed in CAD/CAM programs are exported with the ".stl" (stereolithography) extension and converted into a valuable product in 3D printers (Berman, 2012; Ligon et al., 2017). In order to give the final shape to the product, other post-production processes can be applied according to the 3D printer technology and material used or according to the desired product feature (such as coloring, polishing, strength enhancement, surface smoothing, and stress relief) (Şüheda et al., 2020).

In recent years, there has been an increase in the variety of materials utilized for 3D printing as a result of ongoing research and development efforts. Thermoplastic polymers such as polycarbonate (PC), acrylonitrile-butadiene-styrene (ABS), poly(phenylene sulfone) (PPSU), polyhydroxyalkanoate (PHA), PC-ABS blends, polylactic acid (PLA), polyamide (PA), polypropylene (PP), and polystyrene (PS) are commonly employed as filament materials. Additionally, other materials such as ultraviolet (UV) cured resins, waxes, polymers in foam and powder forms, carbon fiber, aluminum (Al)-infused PA, stainless steel, alloys containing Al, copper (Cu), iron (Fe), nickel (Ni), and cobalt (Co), ceramic powders, silica, sand, graphite, and gypsum materials are also utilized in the 3D printing process (Pop et al., 2019; Valino et al.,2019; Meyer et al., 2020).

Table 3. Categorized 3D printing technology, processing, and their application areas

Classification	Technology	Materials	Process	Applications
Material extrusion	Fused deposition modeling (FDM)/Fused filament fabrication (FFF)	Thermoplastic polymer, wax	Builds objects by dispensing melted thermoplastic (filament) through a nozzle to construct layers	Prototypes, electrical housings, form and fit testings, jigs and fixtures, investment casting patterns, houses
VAT photopolymerization	Stereolithography (SLA) Digital light processing (DLP) Liquid crystal display (LCD)	Photopolymer, ceramic	Creates parts by curing liquid photopolymer (resin) in a vat of photopolymer	Injection molds, polymer prototypes and end-use parts, jewelry industry, dental applications, consumer products
Material jetting	Material Jetting (MJ), NanoParticle Jetting (NPJ)	Photopolymer, wax	Creates parts by depositing droplets of liquid photosensitive fusing agents, followed by curing with light	Full-color product prototypes, injection mold-like prototypes, low-run injection molds, medical models, fashion
Binder jetting	Polymer binder jetting 3D printing Ink-jetting	Metal, polymer, ceramic	Builds parts by adding a binding agent to join powdered material and sintering	Functional metal parts, full-color models, sand casts, and molds
Powder bed fusion	Selective laser sintering (SLS) Selective laser melting (SLM) Direct metal laser sintering (DMLS)	Metal, polymer, ceramic	It builds samples using a high-energy source to fuse regions of powder particles	Functional parts, complex ducting (hollow designs), low-run part production
Direct energy deposition	Electron beam additive melting (EBAM)	Metal	Gives parts by using thermal energy to be deposited and fused simultaneously	Repair of high-end automotive/aerospace components, functional prototypes, and final parts
Sheet lamination	Laminated object manufacture (LOM)	Thermoplastic polymer, composite, paper	Creates parts by trimming individual sheets of material and laminating them into layers	Nonfunctional prototypes, multi-color prints, casting molds

3D printers can produce objects using various materials available in various forms, such as powder, filament, pellets, granules, resin, and more. Table 3 illustrates the different types of 3D printing technologies that utilize these materials and their processing techniques and application areas (Stansbury and Idacavage, 2016; Gross et al., 2017; Beyhan and Arslan Selçuk, 2018; Karagöz et al., 2021). Currently, many 3D printing materials are readily accessible for practical use. These materials find frequent application in diverse sectors such as aviation, automotive, aerospace, oceanography, dentistry, medicine, medical practices, biological systems, molding applications, nanocomposite production, energy sector, construction, textile, electronics, military, food industry, education, and art and hobby applications (Oberoi et al., 2018; Dizon et al., 2020; Cakir Yigit and Karagoz, 2023).

FDM offers the most significant advantages for producing thermoplastic components among all additive manufacturing processes. There are notable advantages of 3D printing, such as a low initial investment, minimal waste, quick material interchange, and commercial software availability. This process enables efficient production of thermoplastic polymer components, especially those with low stiffness requirements (Ngo et al., 2018).

In addition to these unique advantages of 3D printing, there are also some disadvantages and difficulties. Some of the disadvantages of 3D technologies are the mechanical anisotropy in the parts produced with the 3D printing, inhomogeneous microstructure, internal stresses that occur due to heating and cooling during the production of the part, and the effect the part performance, the need for additional surface treatment on the part surfaces after production in some 3D methods, the critical processing speed. Depending on the technology used in the 3D method, some limitations still need to be fully resolved. These limitations are; (i) limited options in terms of material, color, and surface properties, (ii) the low strength of the manufactured parts in terms of temperature, humidity, and brittleness, (iii) The cost of the part increases exponentially as the size of the part to be produced in the 3D printer grows, (iv) lower dimensional accuracy compared to other production methods such as machining, and difficulty in producing in original dimensions, (v) high 3D printer prices, (vi) the possibility of unauthorized reproduction of CAD files/designs (Yıldırım et al., 2018; Karakurt and Lin, 2020).

3D Printing and Injection Molding

Two of the most basic polymer processing technologies today are injection molding and 3D printing. However, they are distinguished by differences in certain operations, capabilities, and applications, but each manufacturing process has advantages and can complement each other. The injection molding process is more efficient for high-volume production with minimal material waste. 3D printing is a slower process but faster to set up, allows for complex design changes, and is better for producing complicated geometry, as depicted in Table 4 (Berman, 2012; Jahan and El-Mounayri, 2018; Kaynak and Varsavas, 2019; Fuenmayor et al., 2019).

Table 4. Comparison of injection molding and 3D printing

	Injection molding	3D printing
Advantages	• Uses multiple molds simultaneously, which means it is more cost-effective to produce many objects • Suitable for dense materials like concrete • Ideal for cost-effective mass production of objects	• The cost of a desktop 3D printer and materials is low • Allows for design changes during production, saving time and money • Ideal for manufacturing intricate or complex designs with intricate internal structures
Disadvantages	• The presence of a mold imposes design restrictions on this production method • Challenging to rectify errors or make alterations to designs • Tailored for industrial use and unsuitable for hobby use as the entry costs are expensive	• Although the setup time is relatively fast, it is a slow production method • Restricted by the printing area size, resulting in the inability to create larger items • The rough surface finish of 3D printed parts can be attributed to the additive layering process

There is generally more crystallinity in thermoplastic components produced by the FDM process than in injection-molded components. Various factors determine the quality of FDM specimens, including the infill pattern and density, layer thickness and width, the air gap between layers, extruder speed and temperature, build orientation, and nozzle diameter. It is vital to investigate the effect of FDM process parameters on specimens' mechanical and thermal properties (Shojib Hossain et al., 2013; Song et al., 2017). FDM specimens of ABS polymer were investigated for their tensile, Izod impact, and hardness properties by Shubham et al. They conducted a comparison with samples manufactured using a conventional injection molding process. The findings revealed that the properties of the injection-molded samples were slightly superior (Shubham et al., 2016). A study by Priya et al. examined the influence of layer heights and infill patterns on the mechanical, thermal, and crystallographic properties of 3D-printed ABS and PLA polymers. The Arrhenius model was used to investigate the thermal behavior of these thermoplastic polymers. Their findings indicated that the data would be helpful to designers in creating numerical models based on finite element analysis (FEA) for thermoplastic polymers (Priya et al., 2019).

Ecker et al. studied how water absorption affects the mechanical properties, impact strength, and morphology of PLA and PLA/wood composites obtained by injection molding and 3D printing. As expected, the mechanical properties of the 3D-printed samples were considerably lower compared to the injection-molded parts. However, in both the injection-molded and 3D-printed samples, the mechanical properties experienced significant reduction following water storage. Consequently, this factor may restrict their potential application range (Ecker et al., 2019). Cisneros-López et al. compared the thermomechanical properties and surface morphology of biocomposites made from recycled post-industrial PLA from coffee pods with specimens 3D printed and injection molded. The thermal properties of both processing methods were very similar. However, due to porosity, 3D printed samples' mechanical performance could have been better than injection molded samples. Finally, they concluded that FDM-based 3D printing techniques can produce complex-shaped biocomposites. They noted that more work should be done to improve the properties of sustainable 3D printed materials (Cisneros-López et al., 2020). Komal et al. recently compared the performance of PLA parts designed by 3D Printing and injection molding. The findings revealed that the printed specimens exhibited superior thermal stability in comparison to the molded specimens. However, no significant differences occurred in the crystallinity of the printed and molded specimens. Furthermore, they concluded that 3D-printed specimens had better tensile, flexural, and thermal characteristics than injection-molded specimens when optimized process parameters (Komal et al., 2021).

Both processes have advantages and disadvantages, making viewing them as complementary technologies more appropriate than competing ones. Traditionally, injection molds are manufactured using machining methods, which can be time-consuming and costly in labor, time, and materials. On the other hand, machined metal molds are durable and suitable for high-volume part production. However, when specific part demands or low production requirements with frequent design changes are necessary, the mold may lose its value once the demand is satisfied (Boros et al., 2019; Dizon et al., 2019). Recently, advancements in 3D printing have made it possible to produce injection molds using additive manufacturing techniques. 3D printing allows for the creation of intricate mold geometries, including internal features and cooling channels, thereby enhancing the efficiency and quality of the injection molding process (Lozano et al., 2022). For instance, Altaf et al. compared the effectiveness of polymer molds created using the FDM 3D printing process with an aluminum mold for possible application in the metal injection molding (MIM) process. The findings revealed that, for a limited number of MIM cycles, parts produced using ABS and nylon mold inserts demonstrated satisfactory performance similar to that of the machine metal mold. Consequently, employing enhanced polymer mold inserts could be a feasible choice in the MIM process for producing low-volume parts, creating prototypes, validating designs, shaping forms, and conducting other preliminary processes before permanent mold production (Altaf et al., 2018).

Several researchers have employed genetic algorithms to optimize the cooling systems, enabling the release of trapped air within the molds. This optimization enhances the injected product's quality, heat transfer, channel geometry, and formability (Yuan et al., 2019). Moreover, rapid prototyping technologies have been utilized to fabricate molds with low-pressure cooling channels for various materials. Recent studies in this field have concentrated on minimizing cooling durations (Kuo and Xu, 2018; Zhang et al., 2018). Conformal cooling channels (CCC) exhibit considerable promise as a viable alternative to conventional straight-drilled cooling channels. The product quality and efficiency are significantly improved since they provide superior cooling effectiveness and uniformity (Feng et al., 2021). CCC designs effectively decrease the cooling time, and the extent of this reduction depends on the shape of the mold's geometry. These designs offer improved surface temperature uniformity, reduced distortions, and consistent cooling performance. These enhancements contribute to CCC's increased efficiency and consistency compared to conventional channels (Kanbur et al., 2020).

Papadakis et al. introduced a method in which mold inserts with conformal cooling channels were produced using selective laser melting (SLM) additive manufacturing techniques to decrease process cycles. Furthermore, numerical analysis was performed to investigate the injection process and simulate shape distortions following the SLM process. They fabricated the injection-molding process setup, as shown in Figure 4. The Figure shows the material input process in an injection machine (top left). At the same time, the right portion demonstrates the injection mold system, consisting of the fixed plate accommodating the mold inserts cavity and the movable platform with the mold thread. The researchers determined the filling pressure to minimize the process pressure, thereby avoiding mold fatigue and damage to the final molded product. As a result of the introduction of the new method for injection molded thermoplastics, cycle times have shortened by up to 32%, and the shape quality of the final product has improved by up to 77% when mold inserts with conformal cooling channels are used over traditional mold inserts (Papadakis et al., 2020).

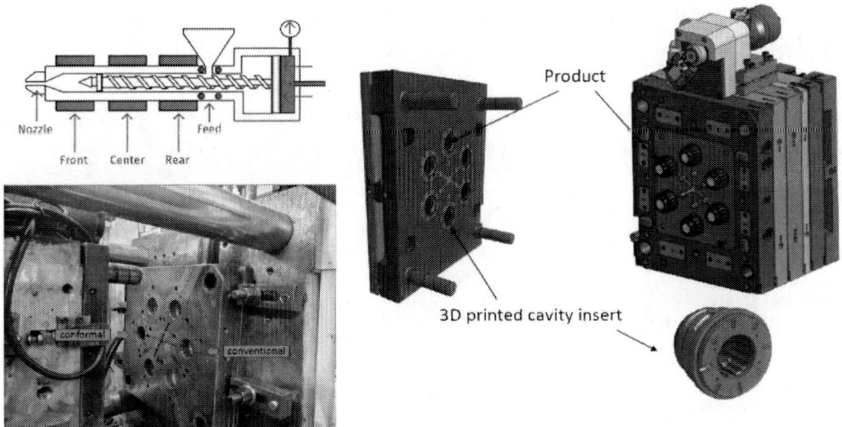

Figure 4. A schematic diagram of the input process (left), the injection mold system (right) with a fixed platform with the mold inserts and polymer products, and a movable platform with the mold threads, along with Elysee Irrigation Ltd.'s injection molding machine setup using convectional and conformal inserts alternately on the six series production mold (adapted with permission). Copyright 2020, MDPI.

Conclusion

Injection molding and 3D printing are important technologies in the manufacturing industry, each with unique advantages and applications. Injection molding is known for its high-quality, large-scale production, providing durable parts with precise dimensions and surface quality. On the other hand, 3D printing is ideal for customized production, allowing the creation of parts in various shapes and sizes, and it is widely used in prototyping.

Both technologies have seen significant advancements in materials, design features, and production methods, expanding their usability. This study successfully compares their advantages and applications, offering valuable insights to industry professionals and researchers seeking efficient utilization of these technologies. The ongoing evolution of manufacturing processes has led to a convergence between injection molding and 3D printing. Advanced 3D printers can now directly produce parts using materials similar to those in injection molding, potentially reducing costs and accelerating the transition from prototyping to mass production. In conclusion, injection molding and 3D printing are essential technologies in part production within the manufacturing industry, each with distinct advantages and applications. This study serves as a valuable resource for professionals seeking effective utilization of these technologies in the design-to-product conversion process. Embracing the possibilities offered by these technologies will undoubtedly shape the future of manufacturing and foster more innovative and flexible approaches in the industry.

Disclaimer

None.

References

Akkurt S. Plastik Malzeme Bilimi Teknolojisi ve Kalıp Tasarımı. *Birsen Yayın Evi* (2007) İstanbul, Türkiye.

Altaf, K, Qayyum, J, Rani, A, Ahmad, F, Megat-Yusoff, P, Baharom, M, Aziz, A, Jahanzaib, M, & German, R. Performance analysis of enhanced 3D printed polymer molds for metal injection molding process. *Metals* (2018) 8(6): 433.

Bendada, A, Derdouri, A, Lamontagne, M, & Simard, Y. Analysis of thermal contact resistance between polymer and mold in injection molding. *Applied thermal engineering* (2004) 24(14-15): 2029-2040.

Berman B. 3-D printing: The new industrial revolution. *Business Horizons* (2012) 55(2): 155-162.

Beyhan F, Arslan Selçuk S. 3D printing in architecture: One step closer to a sustainable built environment. *In Proceedings of 3rd International Sustainable Buildings Symposium (ISBS 2017)* (2018) Volume 1 3 (pp. 253-268). Springer International Publishing.

Blok, LG, Longana, ML, Yu, H, & Woods, BKS. An investigation into 3D printing of fibre reinforced thermoplastic composites. *Additive Manufacturing* (2018) 22: 176-186.

Boros R, Rajamani PK, Kovács JG. Combination of 3D printing and injection molding: Overmolding and overprinting. *Express Polymer Letters* (2019) 13(10): 889-897.

Cakir Yigit N, Karagoz I. A review of recent advances in bio-based polymer composite filaments for 3D printing. *Polymer-Plastics Technology and Materials* (2023) 62(9): 1077-1095.

Cisneros-López, EO, Pal, AK, Rodriguez, AU, Wu, F, Misra, M, Mielewski, DF, Kiziltas, A, & Mohanty, AK. Recycled poly (lactic acid)–based 3D printed sustainable biocomposites: a comparative study with injection molding. *Materials Today Sustainability* (2020) 7: 100027.

Dizon, JRC, Valino, AD, Souza, LR, Espera, AH, Chen, Q, & Advincula, RC. 3D printed injection molds using various 3D printing technologies. In: *Materials Science Forum*. Trans Tech Publications Ltd (2020) 150-156.

Dizon, JRC, Valino, AD, Souza, LR, Espera, AH, Chen, Q, & Advincula, RC. Three-dimensional-printed molds and materials for injection molding and rapid tooling applications. *MRS Communications* (2019) 9(4): 1267-1283.

Ecker, JV, Haider, A, Burzic, I, Huber, A, Eder, G, & Hild, S. Mechanical properties and water absorption behaviour of PLA and PLA/wood composites prepared by 3D printing and injection moulding. *Rapid Prototyping Journal* (2019) 25(4): 672-678.

Farotti E, Natalini M. Injection molding. Influence of process parameters on mechanical properties of polypropylene polymer. A first study. *Procedia Structural Integrity* (2018) 8: 256-264.

Feng S, Kamat AM, Pei Y. Design and fabrication of conformal cooling channels in molds: Review and progress updates. *International Journal of Heat and Mass Transfer* (2021) 171: 121082.

Fu H, Xu H, Liu Y, Yang Z, Kormakov S, Wu D and Sun J. Overview of injection molding technology for processing polymers and their composites. *ES Materials & Manufacturing* (2020) 8(4): 3-23.

Fuenmayor, E, O'Donnell, C, Gately, N, Doran, P, Devine, DM, Lyons, JG, McConville, C, & Major, I. Mass-customization of oral tablets via the combination of 3D printing and injection molding. *International journal of pharmaceutics* (2019) 569: 118611.

Giboz J, Copponnex T, Mélé P. Microinjection molding of thermoplastic polymers: morphological comparison with conventional injection molding. *Journal of Micromechanics and Microengineering* (2009) 19(2): 025023.

Gross B, Lockwood SY, Spence DM. Recent advances in analytical chemistry by 3D printing. *Analytical chemistry* (2017) 89(1): 57-70.

Heckele M, Schomburg WK. Review on micro molding of thermoplastic polymers. *Journal of Micromechanics and Microengineering* (2003) 14(3): R1-R14.

Jahan SA, El-Mounayri H. A thermomechanical analysis of conformal cooling channels in 3D printed plastic injection molds. *Applied Sciences* (2018) 8(12): 2567.

Jeong, E, Kim, Y, Hong, S, Yoon, K, & Lee, S. Innovative injection molding process for the fabrication of woven fabric reinforced thermoplastic composites. *Polymers* (2022) 14(8): 1577.

Kanbur BB, Suping S, Duan F. Design and optimization of conformal cooling channels for injection molding: a review. *The International Journal of Advanced Manufacturing Technology* (2020) 106: 3253-3271.

Kandemir M, Karagöz İ, Sepetçioğlu H. Experimental Investigation of Effects of the Nucleating Agent on Mechanical and Crystallization Behavior of Injection-Molded Isotactic Polypropylene. *El-Cezeri* (2023) 10(1): 109-120.

Karagöz İ, Danış Bekdemir A, Özlem T. 3B Yazıcı Teknolojilerindeki Kullanılan Yöntemler ve Gelişmeler Üzerine Bir Derleme. *Düzce Üniversitesi Bilim ve Teknoloji Dergisi* (2021) 9(4): 1186-1213.

Karagöz İ, Tuna Ö. Effect of melt temperature on product properties of injection-molded high-density polyethylene. *Polymer Bulletin* (2021) 78:6073-6091.

Karagöz İ. An effect of mold surface temperature on final product properties in the injection molding of high-density polyethylene materials. *Polymer Bulletin* (2021) 78:2627-2644.

Karagöz İ. Bilgisayar destekli programlar kullanılarak hazırlanmış döküm kalıbı ve ürün tasarımının polimer kompozit malzemeden üretilmesi. *El-Cezeri* (2018) 5(2): 346-352.

Karakurt I, and Lin L. 3D printing technologies: techniques, materials, and post-processing. *Current Opinion in Chemical Engineering* (2020) 28: 134-143.

Kaynak C, Varsavas SD. Performance comparison of the 3D-printed and injection-molded PLA and its elastomer blend and fiber composites. *Journal of Thermoplastic Composite Materials* (2019) 32(4): 501-520.

Komal UK, Kasaudhan BK, Singh I. Comparative performance analysis of polylactic acid parts fabricated by 3D printing and injection molding. *Journal of Materials Engineering and Performance* (2021) 30(9): 6522-6528.

Kuo CC, Xu WC. Effects of different cooling channels on the cooling efficiency in the wax injection molding process. *The International Journal of Advanced Manufacturing Technology* (2018) 98: 887-895.

Ligon, SC, Liska, R, Stampfl, J, Gurr, M, & Mülhaupt, R. Polymers for 3D printing and customized additive manufacturing. *Chemical Reviews* (2017) 117(15): 10212-10290.

Liou AC, Chen RH. Injection molding of polymer micro-and sub-micron structures with high-aspect ratios. *The International Journal of Advanced Manufacturing Technology* (2006) 28: 1097-1103.

Lozano AB, Álvarez SH, Isaza CV, Montealegre-Rubio W. Analysis and advances in additive manufacturing as a new technology to make polymer injection molds for world-class production systems. *Polymers* (2022) 14(9): 1646.

Massé, H, Arquis, É, Delaunay, D, Quilliet, S, & Le Bot, PH. Heat transfer with mechanically driven thermal contact resistance at the polymer–mold interface in injection molding of polymers. *International Journal of Heat and Mass Transfer* (2004) 47(8-9): 2015-2027.

Meyer TK, Tanikella NG, Reich MJ, Pearce JM. Potential of distributed recycling from hybrid manufacturing of 3-D printing and injection molding of stamp sand and acrylonitrile styrene acrylate waste composite. *Sustainable Materials and Technologies* (2020) 25: e00169.

Ngo TD, Kashani A, Imbalzano G, Nguyen, KTQ, & Hui, D. Additive manufacturing (3D printing): A review of materials, methods, applications and challenges. *Composites Part B: Engineering* (2018) 143: 172-196.

Oberoi, G, Nitsch, S, Edelmayer, M, Janjić, K, Müller, AS, & Agis, H. 3D Printing—encompassing the facets of dentistry. *Frontiers in Bioengineering and Biotechnolog* (2018) 6: 172.

Pantani, R, Coccorullo, I, Speranza, V, & Titomanlio, G. Modeling of morphology evolution in the injection molding process of thermoplastic polymers. *Progress in polymer science* (2005) 30(12): 1185-1222.

Papadakis L, Avraam S, Photiou D, Masurtschak S, and Pereira Falcón JC. Use of a holistic design and manufacturing approach to implement optimized additively manufactured mould inserts for the production of injection-moulded thermoplastics. *Journal of Manufacturing and Materials Processing* (2020) 4(4): 100.

Pop, MA, Croitoru, C, Bedő, T, Geamăn, V, Radomir, I, Cosnită, M, Zaharia, SM, Chicos, LA, & Milosan, I. Structural changes during 3D printing of bioderived and synthetic thermoplastic materials. *Journal of Applied Polymer Science* (2019) 136(17): 47382.

Selva Priya, M, Naresh, K, Jayaganthan, R, & Velmurugan, R. A comparative study between in-house 3D printed and injection molded ABS and PLA polymers for low-frequency applications. *Materials Research Express* (2019) 6(8): 085345.

Saçak M. Polimer Teknolojisi. *Gazi Kitapevi* (2014), Ankara, Türkiye.

Santis, FD, Pantani, R, Speranza, V, & Titomanlio, G. Analysis of shrinkage development of a semicrystalline polymer during injection molding. *Industrial & Engineering Chemistry Research* (2010) 49(5): 2469-2476.

Schönhoff, LM, Mayinger, F, Eichberger, M, Reznikova, E, & Stawarczyk, B. 3D printing of dental restorations: Mechanical properties of thermoplastic polymer materials. *Journal of the Mechanical Behavior of Biomedical Materials* (2021) 119: 104544.

Shojib Hossain M, Ramos J, Espalin D, Perez M, Wicker R. Improving tensile mechanical properties of FDM-manufactured specimens via modifying build parameters. *2013 International Solid Freeform Fabrication Symposium.* (2013) University of Texas at Austin.

Shubham P, Sikidar A, Chand T. The influence of layer thickness on mechanical properties of the 3D printed ABS polymer by fused deposition modeling. *Key engineering materials* (2016) 706: 63-67.

Song, Y, Li, Y, Song, W, Yee, K, Lee, K-Y, & Tagarielli, VL. Measurements of the mechanical response of unidirectional 3D-printed PLA. *Materials & Design* (2017) 123: 154-164.

Stansbury JW, Idacavage MJ. 3D printing with polymers: Challenges among expanding options and opportunities. *Dental Materials* (2016) 32(1): 54-64.

Şüheda Ö, Zeren M, Alp NÇ. 3D yazicilar ile katmanli imalat teknolojisinin otomotiv endüstrisinde uygulanmasi. *International Journal of 3D Printing Technologies and Digital Industry* (2020) 4(1): 18-31.

Sun, X, Zeng, D, Tibbenham, P, Su, X, & Kang, H. A new characterizing method for warpage measurement of injection-molded thermoplastics. *Polymer Testing* (2019) 76: 320-325.

Valino, AD, Dizon, JRC, Espera, AH, Chen, Q, Messman, J, & Advincula, RC. Advances in 3D printing of thermoplastic polymer composites and nanocomposites. *Progress in Polymer Science* (2019) 98: 101162.

Wang, J, Hopmann, C, Kahve, C, Hohlweck, T, & Alms, J. Measurement of specific volume of polymers under simulated injection molding processes. *Materials & Design* (2020) 196: 109136.

Wilczyński K, Wilczyński KJ, Buziak K Modeling and experimental studies on polymer melting and flow in injection molding. *Polymers* (2022) 14(10): 2106.

Xu Y, Zhang Q, Zhang W, Zhang P. Optimization of injection molding process parameters to improve the mechanical performance of polymer product against impact. *The International Journal of Advanced Manufacturing Technology* (2015) 76: 2199-2208.

Yıldırım G, Yıldırım S, Çelik E. Yeni bir bakış - 3 boyutlu yazıcılar ve öğretimsel kullanımı: Bir içerik analizi. *Bayburt Eğitim Fakültesi Dergisi* (2018) 13(25): 163-184.

Yuan Z, Wang H, Wei X, Yan, K, & Gao, C. Multiobjective optimization method for polymer injection molding based on a genetic algorithm. *Advances in Polymer Technology* (2019).

Zhang Y, Hou B, Wang Q, Huang, Z, & Zhou, H. Automatic generation of venting system on complex surfaces of injection mold. *The International Journal of Advanced Manufacturing Technology* (2018) 98: 1379-1389.

Zhao P, Ji K, Zhang J, Chen, Y, Dong, Z, Zheng, J, & Fu, J. In-situ ultrasonic measurement of molten polymers during injection molding. *Journal of Materials Processing Technology* (2021) 293: 117081.

Chapter 3

Injection Molding Conditions for the Production of Bioplastic-Based Products

Mithat Çelebi*
Department of Polymer Materials Engineering, Faculty of Engineering, Yalova University, Yalova, Turkey

Abstract

Injection molding is a popular industrial technique for producing large quantities of plastic parts and test samples. It involves melting a thermoplastic resin and injecting material under high pressure into a mold cavity to get the required shape. Bioplastics have gained popularity in injection molding because of a renewable and sustainable alternative to petroleum-based polymers. Bioplastics are categorized into five classes: biomass derived (polysaccharides and proteins), microbial polyesters (polyhydroxyalkanoates), monomers from sustainable sources (poly(lactic acid), petroleum-derived biodegradable (polycaprolactone, polybutylene adipate terephthalate), and durable bioplastics (bio-based polyethylene, bio-based polypropylene). Injection molding is commonly used to process both bioplastics and conventional polymers. The injection molding processing conditions of each bioplastic class were detailed in this study. The influence of plasticizer and compatibilizer additions in the sample composition on injection molding conditions were explained. In order to increase the quality of bioplastics and lower their prices, injection molding processes and applications of products produced by adding various wastes were discussed. This chapter presents an overview of bioplastics' characteristics, and processing in injection molding.

* Corresponding Author's Email: mithat.celebi@yalova.edu.tr.

In: Injection Molding
Editor: Jose D. Phillips
ISBN: 979-8-89113-429-4

Keywords: bioplastics, injection molding, injection pressure, injection time

Introduction

Injection molding is a popular industrial technique for producing large quantities of complex-shaped plastic parts. It entails melting a thermoplastic resin and injecting it under high pressure into a mold cavity to achieve the required shape (Moritzer and Richters 2021; Ebnesajjad and Khaladkar 2018). Bioplastics have gained popularity in injection molding during the last few decades as a renewable and sustainable alternative to petroleum-based polymers (Ashter 2016).

Bioplastics are classified into the five categories listed below.

Biomass-Derived Bioplastics; Polysaccharides and Proteins

They are obtained from renewable resources (plant and animal based), whey proteins, zein, gelatin and soy proteins. Starch is produced from corn, sugarcane, wheat, or potato starch and modified to increase mechanical and thermal qualities (Alonso-González et al. 2021; Jiménez-Rosado et al. 2022). Cellulose-based bioplastics are produced by chemically modifying plant fibers such as wood, cotton, or hemp to make materials biodegradable (Suryanegara, Nakagaito, and Yano 2009a; 2010; 2009b). They can be composted with organic waste after being degraded by aerobic or anaerobic digestion (Zee 2021)

Microbial Polyesters: Polyhydroxyalkanoates

Polyhydroxyalkanoates (PHAs) are microbial polyesters that can be synthesized by a variety of microbes under nutrient-stress circumstances. They have thermoplastic characteristics similar to those of typical polymers (Müslüm Altun et al. 2023). Bacteria produce PHAs, which can be completely biodegraded into carbon dioxide and water (Vicente, Proença, and Morais 2023).

Monomers Derived from Sustainable Sourced Biodegradable Bioplastics: Poly(Lactic Acid)

Polylactic acid (PLA) is a biodegradable polymer composed of repeating units' lactic acid. Poly(lactic acid) (PLA) is synthesized by different methods using lactic acid produced on a large scale from renewable resources by fermentation (Çelebi and Karagöz 2019; Muslum Altun, Celebi, and Ovali 2022).

Petroleum-Derived Biodegradable Bioplastics: Polycaprolactone, Polybutylene Adipate Terephthalate, etc.

Poly(ε-caprolactone) (PCL) is a semi-crystalline thermoplastic aliphatic polyester with a relatively low melting temperature (60°C) and glass temperature (-60°C) with the repeated unit ε-caprolactone (Çelebi and Karagöz 2019). Poly(butylene adipate-co-terephthalate) (PBAT), a polyester composed of terephthalic acid (TPA), 1,4-butanediol, and adipic acid (Yang et al. 2023).

Durable Bioplastics: Bio-Polyethylene (Bio-PE), Bio-Polypropylene (Bio-PP), etc.

Bio-PE is derived from sugarcane ethanol and can be used in place of fossil-based PE in a variety of applications.

While the first four classes of bioplastics are biodegradable, class 5 bioplastics are not. Biodegradable bioplastics have a lower carbon impact than conventional plastics and are compostable or recyclable (Moshood et al. 2022). Under particular temperature, humidity, and oxygen availability circumstances, they can be broken down into their constituent parts by natural processes such as bacteria or fungi. They must be disposed of in specialist composting facilities since they breakdown at high temperatures and humidity (Silva et al. 2023; Chandra and Rustgi 1998). The monomers of durable bioplastics (class 5) are produced from renewable sources, such as sugar, starch, or vegetable oil, but are not necessarily biodegradable (Atiwesh et al. 2021; Bishop, Styles, and Lens 2021). They can replace petroleum-based plastics in many applications, but proper recycling is required to avoid adverse effects on the environment(Ashter 2016).

Injection Molding of Bioplastics

Bioplastics can be processed using a variety of processes, including compression molding, extrusion, and injection molding. Injection molding of bioplastics conforms to the same basic principles as injection molding of regular plastics, although some modifications are required to adjust the requirements of bioplastics (Santana et al. 2022). Injection molding is a popular manufacturing process that involves melting and shaping polymers (including composites and mixes). This is a useful method for automatically and quickly creating complex forms. Dynamic mold designs can also be customized to make exceedingly complicated designs with the help of robots and standard injection equipment (Harris and Lee 2008).

Bioplastics have lower melting temperatures, viscosity, and shear strength than typical plastics, which can influence molding process parameters like cycle duration, temperature, pressure, and injection speed. Because bioplastics are more susceptible to moisture and can degrade during storage or processing, they require extra care in the drying and handling process (Abe et al. 2021; Folino et al. 2020) Bioplastics are typically combined with additives such as plasticizers, reinforcing agents, and processing aids to obtain optimal molding conditions (Costa et al. 2023; Verbeek and van den Berg 2011). Plasticizers increase the flexibility and toughness of bioplastics, reinforcing agents increase their strength and stiffness, and processing aids reduce viscosity and improve flow. These additives must be carefully chosen to assure compatibility with the bioplastics while minimizing environmental impact (Aguilar et al. 2020).

Injection Molding of Biomass-Derived Bioplastics; Polysaccharides and Proteins

Bagasse, maize cob, cellulose, sago pulp waste, and chitin are examples of agricultural wastes that have the potential to be used as bioplastics. Because of its low cost and ease of availability, farm waste is becoming more popular as a source of bioplastic manufacture. Thermomechanical processing of granular starch results in thermoplastic starch. The processing temperature of thermoplastic starch typically varies from 140 to 160°C (Chan et al. 2021).

Protein-based bioplastics are materials developed from renewable natural components that provide significant benefits due to their technological capacity for a wide range of applications. Furthermore, due to their processability and hydrophilic properties, pea protein isolate, and glycerol appear appropriate to produce novel bioplastics (Carvajal-Piñero et al. 2019). Agricultural food by-products that are high in protein and have a low economic value are employed to produce bioplastics. Studies using various protein-rich byproducts are covered in the literature in this field of research. Felix et al. (2015 and 2016) investigated injection molded crayfish and rice. As injection temperature and pressure, Felix et al. utilized 60-100°C and 0.1-50 MPa, respectively (Félix et al. 2016; Felix et al. 2015). Fernández-Espada et al. (2016), compared injection molded soy and egg white bioplastics. Protein from soybeans is one of the most utilized biopolymer raw materials for bioplastics production since it is a renewable source with a low cost. They investigated various injection pressures (300, 500, and 900 bar) and mold temperatures (60-90°C) (Fernández-Espada et al. 2016). Soy protein is the primary byproduct of soybean oil and is one of nature's cheapest proteins. Jiménez-Rosado et al. (2021) compared the characteristics of soy protein isolate (SPI) and pea protein isolate (PPI) bioplastics treated at various injection times (150, 300, and 450 s) and mold temperatures (70 and 130°C). While a minimum post-injection period was necessary, their findings showed that it has no effect on the mechanical or absorbency qualities of bioplastics. As a result, 150 seconds was a very short time to create bioplastics, resulting in limited mechanical qualities and no water absorption capability. 450 s, on the other hand, had no discernible effect on the characteristics of bioplastics. As a result, 300 s was selected as the best post-injection time (the time required for the bioplastic to develop). SPI bioplastic treated at 70°C will be used if a super absorbent material was required. If greater mechanical strength is desired, the bioplastic can be SPI treated at 130°C. Jiménez-Rosado and colleagues suggested that PPI bioplastics can be employed when a material with low mechanical strength or water absorption capacity was required because this raw material is less expensive than SPI. Processing bioplastics at higher temperatures or pressures resulted in a higher Young's modulus as well as a greater degree of crosslinking. When subjected to a uniaxial tensile load, increasing the mold temperature from 70°C to 130°C leads in greater extensibility of bioplastic samples (Jiménez-Rosado et al. 2021). Perez-Puyana et al. studied the effect of injection moulding processing parameters on the properties of pea protein-based bioplastics (Perez-Puyana et al. 2016). They investigated various molding periods (100, 200, and 300 s) and injection pressures (100, 300, 500, and 900 bar). Long molding times result in lower water absorption, brighter color, and less transparency, whereas short molding times result in higher water absorption, darker color, and higher transparency. Furthermore, the longer molding process promotes generating bioplastics with a more crystalline structure (Perez-Puyana et al. 2016).

Gamero et al. investigated the mechanical and absorption properties of lignocellulose-reinforced bioplastics by injection molding soy protein-based bioplastics with lignocellulosic fibers. They investigated at the impact of mold temperature (70, 90, 110, and 130°C) by using a reasonably high fiber concentration (e.g., 5.0% by weight) in bioplastic formulations and processing at higher temperatures (e.g., 130°C), it was feasible to generate reinforced materials with improved mechanical properties. It exemplified the intriguing role of lignocellulose fiber as an ingredient in protein-based bioplastics (Gamero et al. 2019). Bourny et al. (2017) employed nanoclay (Montmorillonite, MMT-Na) to improve the physicochemical properties of injection molded SPI/MMT nanocomposites(Bourny et al. 2017). Depending on the processing parameters, the addition of MMT-Na+ increased the mechanical characteristics and water holding capacity of Soy protein isolate (SPI) based products. Appropriate processing parameters were chosen based on the findings of temperature ramp and time scan tests conducted in prior investigations using SPI/GL mixes. As a result, the cylinder temperature was set at 40°C and the injection temperature was set at 70°C, which was close to the glassy transition of the mixture. Injection and holding pressures were typically setted to 500 and 900 bar, respectively. Appropriate processing conditions were selected based on the results obtained from the temperature ramp and time scan tests performed by the authors in their previous studies with soy protein isolate/glycerol mixtures. Therefore, the temperature in the cylinder was setted to 40°C and the injection temperature to 70°C, which was close to the glassy transition of the mixture. Injection and holding pressures were generally setted at 500 and 900 bar, respectively. Furthermore, increasing the injection pressure (from 500 bar to 900 bar) improved the tensile characteristics (strain at rupture increases by 100% and maximum stress increases by 52%), which may be attributed to the reduction in nanoclay particle size. However, the injection pressure drop causes only slight variations in water absorption qualities and had no effect on bending properties. Bourny et al. (2017) concluded that increasing injection pressure resulted in a considerable reduction in nanoclay particle size (Bourny et al.

2017). These findings suggested that molding time and injection pressures were critical factors in modifying bioplastic properties. All these authors agree that agri-food by-products were a potential raw material for the creation of competitive and economical bioplastics in the market.

Wittwer and Tomka developed a patent for injection molded medicine capsules in the mid-1980s, which was one of the earliest developments in injection molding of starch products. This product was made from starch and had a moisture content ranging from 5% to 30%. The method was similar to traditional extrusion cooking in that water was used as the principal plasticizer. For injection molding applications, researchers had investigated strategies to mitigate the detrimental consequences of starch's high viscosity and poor flow characteristics. A tried-and-true solution is to combine starch with synthetic polymers, which frequently reduce viscosity (Baltá Calleja et al. 1999).

Injection Molding of Microbial Polyester Bioplastics; Polyhydroxyalkanoates

Polyhydroxyalkanoates (PHAs) are bacterially generated biodegradable and biocompatible polyester compounds (Mazur et al. 2022; Sinha Ray 2013). Injection molding conditions of PHAs are determined by the type of PHAs material used and the product needs (Vanheusden et al. 2021). PHAs are a class of biopolymers that include poly(3-hydroxybutyrate) and its copolymer PHBV (poly(3-hydroxybutyrate-co-3-hydroxyvalate). Rice husk, wheat bran, mango peel, potato peel, pulp, and straw are all agricultural and food wastes that can be utilized to produce the PHAs. PHAs are bacterial intracellular (energy storage) products. PHA can be produced by over 250 different microorganisms (Lackner 2015; Kökpınar and Altun 2022).

Injection Molding of Monomers Derived from Sustainable Sourced Biodegradable Bioplastics: Poly(Lactic Acid)

The development of synthetic polymers based on bio-based monomers provides an alternative to the bioplastics market's expansion. PLA, commonly known as poly(lactic acid), is one of the most well-known polymers in this area. PLA is a biocompatible thermoplastic aliphatic polyester (Lackner 2015).

Injection molding of heat-resistant PLA products necessitates quick crystallization with less than 1% D-isomer, which is commonly achieved with PLA by adding nucleating chemicals. During the rapid cooling cycle in the mold, these compositions allow for the formation of high levels of crystallinity (Chan et al. 2021).

PLA is a hygroscopic thermoplastic that easily absorbs water from the atmosphere. Moisture causes hydrolysis of PLA homopolymer during melt processing, resulting in reduced mechanical performance of the PLA. Even the presence of a small amount of moisture hydrolyzes PLA in the melt phase, reducing the molecular weight and causing loss of properties. Prior to injection molding, all PLA-containing compounds must be dried to a moisture content of less than 0.02% using a desiccant dryer capable of blowing air with a dew point of -40°C. The preferred method for drying PLA is to use a desiccant hot air dryer system. Another option is to use a vacuum drying oven. Injection molding is the process of melting PLA polymer or PLA compound, putting it into a mold under high pressure, and hardening it until a stable product is created. While normal PLA compounds have an amorphous form, high temperature PLA compounds can produce a semi-crystalline structure. Combining PLA and PDLA homopolymers yields high heat PLA. These homopolymers have higher melting points and a faster crystallization rate than regular PLA. As a result, compounds comprising homopolymers are appropriate for the fabrication of semi-crystalline parts with improved temperature resistance. Injection molding recommendations apply to both high-temperature PLA compounds and regular PLA, with a differing mold temperature required. Because injection molding is a broad machining technology with several applications and material formulas, the information in this machining guide is merely a beginning point. Process optimization is recommended to determine the best process conditions for injection molding the intended item. PLA, with a melting temperature of around 175°C, can be processed into molded pieces, films, or fibers using normal plastics processing equipment at temperatures exceeding 185 - 190°C. These procedures (for example, injection molding) subject the polymer to severe heat stress. PLA degrades through hydrolysis, thermal deterioration above 200°C, and other chemical processes. Process time, temperature, low molecular weight contaminants, and catalyst concentration all have an impact on these. As a result, PLA homopolymers have a relatively limited processing window. PLA degradation is accomplished through simple hydrolysis of the ester link and does not require the use of catalysts (e.g., enzymes). Degradation rates are affected by surface area, isomer ratio, and temperature (Tábi, Ageyeva, and Kovács 2022).

The use of poly(lactic acid) (PLA) for injection molded products is limited for commercial applications because PLA has a slow crystallization rate compared to many other thermoplastics besides standard injection molding cycle times. The cycle time must be exceptionally long or a post annealing step must be included to get a component with sufficient crystallinity to maximize physical property improvements during injection molding. Because the average injection molding cycle time is 60-90 seconds, this would be impracticable and economically unfeasible for big volume automobile application (Harris and Lee 2008). Crystallization can improve thermomechanical qualities like as hardness, brittleness, and barrier properties. To improve the crystallinity of polymers, nucleating agent additives are utilized. Injection molding cycle times must be lowered through nucleation. Talc is one of the most effective nucleating agents (Tábi, Ageyeva, and Kovács 2022). PLA's intrinsic fragility and low Heat Deflection Temperature (HDT) make it challenging to use, and both are especially difficult to cure in an injection molded product. At the same time, the properties of PLA can be considerably altered by various means, such as various plasticizers, impact modifiers and blending, copolymerization, and so on. PLA's HDT can be used to boost it. Developed as a result of increasing crystallinity. The incorporation of bio-based reinforcements, resulting in biocomposites, increases both the HDT and overall mechanical performance of PLA (Schäfer, Pretschuh, and Brüggemann 2019).

Scoponi et al. developed bio-based products by combining a star-shaped poly(d,l-lactide) (star-PDLLA) with a typical linear poly(l-lactide) (linear-PLLA). At 190°C, polymer blends of star-PDLLA and linear-PLLA were injection molded. To create a rapid cooling ramp and prevent crystallization, the mold temperature was adjusted to 0°C. Star-PDLLA had shown to be a green and highly compatible plasticizer for brittle linearPLLA due to its low molecular mass and low entanglement structure (Scoponi, Francini, and Athanassiou 2021). The effects of processing parameters, specifically molding temperature, on the mechanical performance of impact modified poly(lactic acid) (PLA) were examined by Ryan Vadori et al. Many of the final properties of the material are affected by polymer crystallization. Increasing the die temperature can boost crystallinity even further. Lowering the temperature of the mold and producing a highly amorphous polymer can increase mechanical qualities like as elongation and impact strength. This is due to the increased amorphous area, which can rearrange the component chains to absorb some of the material's stresses (Vadori, Mohanty, and Misra 2013).

To enhance the flexibility of PLA, Barletta et al. developed the formulation of PLA blends modified with PBAT. The polylactic acid-based blend was produced by injection molding screw caps that can be opened and closed with the addition of another biodegradable polyester, polybutylene adipate-co-terephthalate. The pellets obtained after extrusion were reprocessed in this investigation by sequential injection. During injection molding, the following parameters were used: hot runner temperature 240°C, heated sleeve temperature 185-210°C, mold wall temperature 35°C, injection pressure 700 bar, holding pressure 500 bar, holding time 0.5 sec, cooling time 1.2-1.7 sec, and cycle duration 5-5.5 sec. Prior to injection molding, the compounds were dried for 8 hours in a conventional drier at 60°C. Biodegradable polyester-based blends have been discovered to be appropriate for injection molding of snap-proof screw caps with adequate flexibility during demoulding. Screw caps can also offer excellent temperature stability, mechanical strength, and impact resistance. As a result, combining PLA with PBAT and adequate proportions of processing aids can yield polymeric materials suitable for injection molding biodegradable screw caps (Barletta et al. 2020). Pappu et al. produced hybrid composites from granulated sisal and hemp fibers via extrusion and injection molding with aliphatic polyester (PLA) made of lactic acid and analyzed their performance. Based on mechanical properties, water absorption studies, thermal analysis, crystallinity, and microstructure properties, it is clear that high performance hybrid fiber reinforced composites can be produced using extrusion (melt processing) and injection molding with sisal and hemp fiber in combination with polylactic acid (Pappu, Pickering, and Thakur 2019).

Injection Molding of Petroleum-Derived Biodegradable Bioplastics

Polybutylene adipate-co-terephthalate (PBAT) is a biodegradable polymer with excellent packaging qualities. Pavon et al. utilized pine resin derivatives, gum resin (GR), and the pentaerythritol ester (UT) of GR as sustainable additives to improve the qualities of PBAT and minimize its cost. Melt extrusion followed by injection molding was used to blend PBAT with additions of 5%, 10%, and 15% by weight. The formulations were extruded in a twin-screw extruder at 50 rpm with a temperature profile of 180, 170, 160, and 150°C (die to chamber). The resulting PBAT resin-based materials were pelleted, and injection molded into test specimens in an injection molding machine with a temperature profile from the die to the chamber of 180, 170, 160, 150, and 140°C (Pavon et al. 2020). Andrzejewski et al. used flax fiber to strengthen PLA and PBAT blends. After the material was produced in a twin-screw extruder, injection molding was used to generate samples. Injection molding temperatures were 190°C, mold temperatures were 40°C, injection speed was 100 cm/s, injection/holding pressure was 1050/850 bar, and holding/cooling time was -10/40 s. It is worth noting that it has little impact on the mechanical and thermomechanical properties of injection-molded specimens. The increase in PLA crystalline phase concentration for produced composite samples is primarily due to better processing conditions, particularly a slower cooling rate. In reality, for pure PLA, the increase in crystalline phase concentration varies from 23% (injection molding) to around 41% for composites (Andrzejewski and Nowakowski 2021).

Delgado et al. developed bioplastic materials from rapseed meal and polycaprolactone (PCL) using injection molding at various mold temperatures (80, 100, and 120°C). An increase in injection moulding temperature increased viscoelastic characteristics while decreasing water uptake capacity, implying a direct interaction between protein cross-linking and mould irrespective of meal processing (pelletizing, milling, sieving) (Delgado, Felix, and Bengoechea 2018).

Injection Molding of Durable Bioplastics

Brazilian company (Braskem) was the first to commercialize biobased polyethylene (bio-PE) utilizing local sugar cane-derived ethanol as a raw material. Braskem began commercial production of bio-based High-Density Polyethylene (HDPE) with a capacity of 200,000 tons per year in 2010. The material's composition and performance are equivalent to that of petroleum-based PE (Lackner 2015).

The compatibility of cellulose esters with PLA and biopolyethylene was examined by Willberg-Keyriläinen et al. Testing was done on test specimens utilizing the injection molding technique. Mold temperature for injection molding was 35°C. 5 s of 400 bar for injection pressure and 15 s of 200 bar for holding pressure were employed. The results demonstrated that the investigated thermoplastic cellulose materials had a promising future for usage in injection molding applications (Willberg-Keyriläinen, Orelma, and Ropponen 2018).

Conclusion

Bioplastics, as conventional plastics, are formed via injection molding. Bioplastics are crucial for sustainability since they are made from renewable, biodegradable, and recyclable raw materials. Bioplastics help reduce the environmental impact of plastic waste while supporting the circular economy. Injection of bioplastics demands much more care than injection of ordinary plastics. When the injection temperature and time exceed the optimum conditions, the characteristics of the bioplastic product decrease. These negative impacts can be mitigated by employing various additives.

Plasticization, impact modification mixing, nucleation, stereocomplexing, and in-mold crystallization are used to easy injection molding of PLA. Oligomeric lactic acid as a plasticizer, natural rubber and ethylene vinyl acetate as impact modifiers, and poly(butylene adipate-co-terephthalate) or poly(butylene succinate) as a second and naturally hard biopolymer phase are some of the most acceptable additives for increasing impact resistance of PLA The most crucial finding is that the PLA phase must be crystallized (annealed) to have the greatest impact modifying effect. There are two techniques to raise the HDT of injection molded PLA items. One method is to use a low grade of D-lactide PLA (or PLLA) in combination with effective heterogeneous or stereocomplex (PDLA) nucleating agents to induce crystallization and hence increase crystallinity. Furthermore, the use of nucleating agents should be augmented by post- or in-mold crystallization, which is commonly caused by rapid cooling during injection molding. Another technique to improve HDT is to add chemicals to PLA that considerably increase the hardness of the part (perhaps with some nucleation ability) so that the product can reach a rubbery state at high temperatures without major distortion until cold crystallization occurs. These stiffening additives could be talc or natural plant fibers (biocomposite formulation). However, plant fibers have significant drawbacks, including high natural water content, susceptibility to thermal deterioration, restricted fiber length and fiber content, and flexible fibers' curved shape. Because of this, they are not widely used in injection molding applications (Vadori, Mohanty, and Misra 2013; Pappu, Pickering, and Thakur 2019; Barletta et al. 2020; Scoponi, Francini, and Athanassiou 2021; Tábi, Ageyeva, and Kovács 2022).

Table 1. Injection molding conditions of bioplastics in the literature

	Bioplastic	Additives	Pretreatment	Injection conditions	References
1	Pea Protein Isolate (PPI)	Glycerol (GL)	Time: 60 min. Temperature: 25°C, mix rate: 50 rpm	Injection temp.: 50°C, injection time: 100 s, molding temp.: 130°C, molding time: 100, 200, 300 s, injection pressure: 100-300-500-900 bar	(Perez et al. 2016)
2	Poly(lactic acid) PLA, Bio-PE, Celulose octanate, Cellulose palmitate	LiCl/DMAc	Cellulose was dissolved in 5% LiCl/DMAc. Anhydrous pyridine and octanoyl were added to the cellulose solution. Reaction time: 16 h., 80°C	Mold temp.: 35°C, Injection pressure: 400 bar (5 s), holding pressure: 200 bar (15 s)	(Willberg-Keyriläinen, Orelma, and Ropponen 2018)
4	Bio-HDPE, Pinecone	PE-g-MA	Bio-HDPE and Pinecone powder were dried separately for 48 h., 60°C. The temp. profile was set at 140–145–150– 155°C with a rotational speed of 20 rpm.	Temp. profile in the injection molding unit 140°C (chamber), 150°C, 155°C and 160°C (injection nozzle). 75 tons clamping force, cavity filling and cooling times are 1 and 10 s.	(Morcillo et al. 2021)
5	Bio-PP, Mango Peel (MPF)	PP-g-IA Dicumyl peroxide (DCP)	MPF was dried at 50°C for 48 h. The shells were crushed and ground in a centrifugal mill at 8000 rpm and finally sieved through a 250-mesh mesh sieve to obtain MPF.	The temp. profile in the injection molding unit was 155 °C (chamber), 160, 165 and 170°C (injection nozzle). 75 tons of clamping force applied, cavity filling and cooling times of 1 and 10 s.	(Gomez-Caturla et al. 2022)
6	BioHDPE, hemp, flax, jute	PE-g-MA	Bio-HDPEs were dried in a dehumidification oven at 60°C for 48 hours. fed into the twin screw (L/D: 24) extruder. The extruder was operated at 20 rpm and the temperatures were 140-145-150-155°C.	Injection molding unit: 140°C (chamber), 150°C, 155°C and 160°C (injection nozzle). A closing force of 75 tons. The cavity filling and cooling times were 1s and 10s, respectively.	(Dolza et al. 2021)
7	Bio-HDPE and Micronized Argan Shell (MAS)	PE-g-MA HNT	Bio-HDPE, PE-g-MA, MAS and HNTs were dried at 40°C for 48 hours	Injection machine temperature: 135°C (chamber), 140°C, 150°C and 160°C (injection nozzle) closing force: 75 tons was applied, and the cavity filling and cooling times were set as 1 and 10 s.	(Jorda-Reolid et al. 2021)

More research is needed to address the challenges of bioplastics, such as cost, performance and scalability, to promote their use by industries and consumers. Important parameters in the production of bioplastics by injection molding include mold temperature, injection pressure, holding pressure and cooling times. Other factors affecting bioplastic injection molding conditions include moisture content and the presence of additives. It is very important to strictly adhere to the specifications for the processing of bioplastics and to work within the process ranges recommended by the bioplastic manufacturer.

References

Abe, Mateus Manabu, Júlia Ribeiro Martins, Paula Bertolino Sanvezzo, João Vitor Macedo, Marcia Cristina Branciforti, Peter Halley, Vagner Roberto Botaro, and Michel Brienzo. (2021). "Advantages and Disadvantages of Bioplastics Production from Starch and Lignocellulosic Components." *Polymers*, https://doi.org/10.3390/polym13152484.

Aguilar, José Manuel, Carlos Bengoechea, Eva Pérez, and Antonio Guerrero. (2020). "Effect of Different Polyols as Plasticizers in Soy Based Bioplastics." *Industrial Crops and Products* 153. https://doi.org/10.1016/j.indcrop.2020.112522.

Alonso-González, María, Manuel Felix, Antonio Guerrero, and Alberto Romero. (2021). "Rice Bran-Based Bioplastics: Effects of the Mixing Temperature on Starch Plastification and Final Properties." *International Journal of Biological Macromolecules* 188. https://doi.org/10.1016/j.ijbiomac.2021.08.043.

Altun, Muslum, Mithat Celebi, and Sabih Ovali. (2022). "Preparation of the Pistachio Shell Reinforced PLA Biocomposites: Effect of Filler Treatment and PLA Maleation." *Journal of Thermoplastic Composite Materials*, 35 (9), 1342–57. https://doi.org/10.1177/08927057211010880.

Altun, Müslüm, Mithat Çelebi, Sinan Şen, Öznur Kökpınar, and Hanife Songül Kaçoğlu. (2023). "Characterization of Polyhydroxyalkanoate-Based Composites Derived from Waste Cooking Oil and Agricultural Surplus." *Polymer Composites*, July. https://doi.org/10.1002/pc.27379.

Andrzejewski, Jacek, and Michał Nowakowski. (2021). "Development of Toughened Flax Fiber Reinforced Composites. Modification of Poly(Lactic Acid)/Poly(Butylene Adipate-Co-Terephthalate) Blends by Reactive Extrusion Process." *Materials*,14 (6), https://doi.org/10.3390/ma14061523.

Ashter, Syed Ali. (2016). *Introduction to Bioplastics Engineering. Introduction to Bioplastics Engineering*. https://doi.org/10.1016/B978-0-323-39396-6.00005-1.

Atiwesh, Ghada, Abanoub Mikhael, Christopher C. Parrish, Joseph Banoub, and Tuyet Anh T. Le. (2021). "Environmental Impact of Bioplastic Use: A Review." *Heliyon*, 7 (9), https://doi.org/10.1016/j.heliyon.2021.e07918.

Baltá Calleja, F. J., Rueda, D. R., Secall, T., Bayer, R. K., and Schlimmer, M. (1999). "Influence of Processing Methods on Starch Properties." *Journal of Macromolecular Science – Physics*, 38, https://doi.org/10.1080/00222349908212444.

Barletta, Massimiliano, Clizia Aversa, Michela Puopolo, and Silvia Vesco. (2020). "Ultra-Flexible PLA-Based Blends for the Manufacturing of Biodegradable Tamper-Evident Screw Caps by Injection Molding." *Journal of Applied Polymer Science*, 137 (46), https://doi.org/10.1002/app.49428.

Bishop, George, David Styles, and Piet N. L. Lens. (2021). "Environmental Performance Comparison of Bioplastics and Petrochemical Plastics: A Review of Life Cycle Assessment (LCA) Methodological Decisions." *Resources, Conservation and Recycling*. https://doi.org/10.1016/j.resconrec.2021.105451.

Bourny, V., Perez-Puyana, V., Felix, M., Romero, A., and Guerrero, A. (2017). "Evaluation of the Injection Moulding Conditions in Soy/Nanoclay Based Composites." *European Polymer Journal*, 95 (April 2017), 539–46. https://doi.org/10.1016/j.eurpolymj.2017.08.036.

Carvajal-Piñero, J. M., Ramos, M., Jiménez-Rosado, M., Perez-Puyana, V., and Romero, A. (2019). "Development of Pea Protein Bioplastics by a Thermomoulding Process: Effect of the Mixing Stage." *Journal of Polymers and the Environment*, 27 (5), 968–78. https://doi.org/10.1007/s10924-019-01404-3.

Çelebi, Mithat, and İdris Karagöz. (2019). "Biyobozunur Polimerler Ve Özellikleri; Nişasta, Poli(Glikolik Asit), Poli(Laktik Asit) Ve Poli(Ɛ Kaprolakton)." In *Mühendislik Alanında Yeni Ufuklar*, edited by Mahmut Turhan, 275–92. Ankara: Gece Kitabevi. www.gecekitapligi.com.

Chan, Jia Xin, Joon Fatt Wong, Azman Hassan, and Zainoha Zakaria. (2021). "Bioplastics from Agricultural Waste." In *Biopolymers and Biocomposites from Agro-Waste for Packaging Applications*. https://doi.org/10.1016/b978-0-12-819953-4.00005-7.

Chandra, R, and Renu Rustgi. (1998). "Pergamon BIODEGRADABLE POLYMERS." *Progress in Polymer Science*, 23 (97), 1273–1335. https://doi.org/10.1016/S0079-6700(97)00039-7.

Costa, Ana, Telma Encarnação, Rafael Tavares, Tiago Todo Bom, and Artur Mateus. (2023). "Bioplastics: Innovation for Green Transition." *Polymers*. https://doi.org/10.3390/polym15030517.

Delgado, M., M. Felix, and C. Bengoechea. 2018. "Development of Bioplastic Materials: From Rapeseed Oil Industry by Products to Added-Value Biodegradable Biocomposite Materials." *Industrial Crops and Products*, 125, https://doi.org/10.1016/j.indcrop.2018.09.013.

Dolza, Celia, Eduardo Fages, Eloi Gonga, Jaume Gomez-Caturla, Rafael Balart, and Luis Quiles-Carrillo. (2021). "Development and Characterization of Environmentally Friendly Wood Plastic Composites from Biobased Polyethylene and Short Natural Fibers Processed by Injection Moulding." *Polymers*, 13 (11), https://doi.org/10.3390/polym13111692.

Ebnesajjad, Sina, and Pradip R. Khaladkar. (2018). "Manufacturing Parts From Melt-Processible Fluoropolymers." In *Fluoropolymer Applications in the Chemical Processing Industries*. https://doi.org/10.1016/b978-0-323-44716-4.00006-3.

Félix, M., Lucio-Villegas, A., Romero, A., and Guerrero, A. (2016). "Development of Rice Protein Bio-Based Plastic Materials Processed by Injection Molding." *Industrial Crops and Products*, 79, https://doi.org/10.1016/j.indcrop.2015.11.028.

Felix, Manuel, Alberto Romero, Felipe Cordobes, and Antonio Guerrero. (2015). "Development of Crayfish Bio-Based Plastic Materials Processed by Small-Scale Injection Moulding." *Journal of the Science of Food and Agriculture*, 95 (4), https://doi.org/10.1002/jsfa.6747.

Fernández-Espada, Lucía, Carlos Bengoechea, Felipe Cordobés, and Antonio Guerrero. (2016). "Thermomechanical Properties and Water Uptake Capacity of Soy Protein-Based Bioplastics Processed by Injection Molding." *Journal of Applied Polymer Science*, 133 (24), https://doi.org/10.1002/app.43524.

Folino, Adele, Aimilia Karageorgiou, Paolo S. Calabrò, and Dimitrios Komilis. (2020). "Biodegradation of Wasted Bioplastics in Natural and Industrial Environments: A Review." *Sustainability (Switzerland)*. https://doi.org/10.3390/su12156030.

Gamero, S., Jiménez-Rosado, M., Romero, A., Bengoechea, C., and Guerrero, A. (2019). "Reinforcement of Soy Protein-Based Bioplastics Through Addition of Lignocellulose and Injection Molding Processing Conditions." *Journal of Polymers and the Environment*. https://doi.org/10.1007/s10924-019-01430-1.

Gomez-Caturla, Jaume, Rafael Balart, Juan Ivorra-Martinez, Daniel Garcia-Garcia, Franco Dominici, Debora Puglia, and Luigi Torre. (2022). "Biopolypropylene-Based Wood Plastic Composites Reinforced with Mango Peel Flour and Compatibilized with an Environmentally Friendly Copolymer from Itaconic Acid." *ACS Applied Polymer Materials*, 4 (6), https://doi.org/10.1021/acsapm.2c00373.

Harris, Angela M., and Ellen C. Lee. (2008). "Improving Mechanical Performance of Injection Molded PLA by Controlling Crystallinity." *Journal of Applied Polymer Science*, 107 (4), https://doi.org/10.1002/app.27261.

Jiménez-Rosado, Mercedes, Jean Eudes Maigret, Denis Lourdin, Antonio Guerrero, and Alberto Romero. (2022). "Injection Molding versus Extrusion in the Manufacturing of Soy Protein-Based Bioplastics with Zinc Incorporated." *Journal of Applied Polymer Science*, 139 (7), https://doi.org/10.1002/app.51630.

Jiménez-Rosado, Mercedes, Jose Fernando Rubio-Valle, Víctor Perez-Puyana, Antonio Guerrero, and Alberto Romero. (2021). "Comparison between Pea and Soy Protein-Based Bioplastics Obtained by Injection Molding." *Journal of Applied Polymer Science*, 138 (20), https://doi.org/10.1002/app.50412.

Jorda-Reolid, Maria, Jaume Gomez-Caturla, Juan Ivorra-Martinez, Pablo Marcelo Stefani, Sandra Rojas-Lema, and Luis Quiles-Carrillo. (2021). "Upgrading Argan Shell Wastes in Wood Plastic Composites with Biobased Polyethylene Matrix and Different Compatibilizers." *Polymers*, 13 (6), https://doi.org/10.3390/polym13060922.

Kökpınar, Öznur, and Müslüm Altun. (2022). "Evaluation of Different Nutrient Limitation Strategies for the Efficient Production of Poly(Hydroxybutyrate-co-hydroxyvalerate) from Waste Frying Oil and Propionic Acid in High Cell Density Fermentations of Cupriavidus Necator H16." *Preparative Biochemistry and Biotechnology*, 0 (0), 1–10. https://doi.org/10.1080/10826068.2022.2114009.

Lackner, Maximillian. (2015). *Bioplastics - Biobased Plastics as Renewable and/or Biodegradable Alternatives to Petroplastics. Kirk-Othmer Encyclopedia of Chemical Technology.*

Mazur, Karolina E., Paulina Jakubowska, Anna Gaweł, and Stanisław Kuciel. (2022). "Mechanical, Thermal and Hydrodegradation Behavior of Poly (3-Hydroxybutyrate-Co-3-Hydroxyvalerate) (PHBV) Composites with Agricultural Fibers as Reinforcing Fillers." *Sustainable Materials and Technologies*, 31, https://doi.org/10.1016/j.susmat.2022.e00390.

Morcillo, Maria Del Carmen, Ramón Tejada, Diego Lascano, David Garcia-Sanoguera, and Daniel Garcia-Garcia. (2021). "Manufacturing and Characterization of Environmentally Friendly Wood Plastic Composites Using Pinecone as a Filler into a Bio-Based High-Density Polyethylene Matrix." *Polymers*, 13 (24), https://doi.org/10.3390/polym13244462.

Moritzer, Elmar, and Maximilian Richters. (2021). "Injection Molding of Wood-Filled Thermoplastic Polyurethane." *Journal of Composites Science*, 5 (12), https://doi.org/10.3390/jcs5120316.

Moshood, Taofeeq D., Gusman Nawanir, Fatimah Mahmud, Fazeeda Mohamad, Mohd Hanafiah Ahmad, and Airin AbdulGhani. (2022). "Sustainability of Biodegradable Plastics: New Problem or Solution to Solve the Global Plastic Pollution?" *Current Research in Green and Sustainable Chemistry.* https://doi.org/10.1016/j.crgsc.2022.100273.

Pappu, Asokan, Kim L. Pickering, and Vijay Kumar Thakur. (2019). "Manufacturing and Characterization of Sustainable Hybrid Composites Using Sisal and Hemp Fibres as Reinforcement of Poly (Lactic Acid) via Injection Moulding." *Industrial Crops and Products*, 137, https://doi.org/10.1016/j.indcrop.2019.05.040.

Pavon, Cristina, Miguel Aldas, Harrison de la Rosa-Ramírez, Juan López-Martínez, and Marina P. Arrieta. (2020). "Improvement of Pbat Processability and Mechanical Performance by Blending with Pine Resin Derivatives for Injection Moulding Rigid Packaging with Enhanced Hydrophobicity." *Polymers*, 12 (12), https://doi.org/10.3390/polym12122891.

Perez, Victor, Manuel Felix, Alberto Romero, and Antonio Guerrero. (2016). "Characterization of Pea Protein-Based Bioplastics Processed by Injection Moulding." *Food and Bioproducts Processing*, 97, https://doi.org/10.1016/j.fbp.2015.12.004.

Perez-Puyana, V., Felix, M., Romero, A., and Guerrero, A. (2016). "Effect of the Injection Moulding Processing Conditions on the Development of Pea Protein-Based Bioplastics." *Journal of Applied Polymer Science*, 133 (20), https://doi.org/10.1002/app.43306.

Santana, Ismael, Manuel Félix, Antonio Guerrero, and Carlos Bengoechea. (2022). "Processing and Characterization of Bioplastics from the Invasive Seaweed Rugulopteryx Okamurae." *Polymers*, 14 (2), https://doi.org/10.3390/polym14020355.

Schäfer, Hendrik, Claudia Pretschuh, and Oliver Brüggemann. (2019). "Reduction of Cycle Times in Injection Molding of PLA through Bio-Based Nucleating Agents." *European Polymer Journal*, 115, https://doi.org/10.1016/j.eurpolymj.2019.03.011.

Scoponi, Giulia, Nora Francini, and Athanassia Athanassiou. (2021). "Production of Green Star/Linear PLA Blends by Extrusion and Injection Molding: Tailoring Rheological and Mechanical Performances of Conventional PLA." *Macromolecular Materials and Engineering*, 306 (5), https://doi.org/10.1002/mame.202000805.

Silva, Rafael Resende Assis, Clara Suprani Marques, Tarsila Rodrigues Arruda, Samiris Cocco Teixeira, and Taíla Veloso de Oliveira. (2023). "Biodegradation of Polymers: Stages, Measurement, Standards and Prospects." *Macromol*, 3 (2), https://doi.org/10.3390/macromol3020023.

Sinha Ray, Suprakas. (2013). "Environmentally Friendly Polymer Nanocomposites Using Polymer Matrices from Renewable Sources." In *Environmentally Friendly Polymer Nanocomposites*. https://doi.org/10.1533/9780857097828.1.89.

Suryanegara, Lisman, Antonio Norio Nakagaito, and Hiroyuki Yano. (2009a). "The Effect of Crystallization of PLA on the Thermal and Mechanical Properties of Microfibrillated Cellulose-Reinforced PLA Composites." *Composites Science and Technology*, 69 (7–8), 1187–92. https://doi.org/10.1016/j.compscitech.2009.02.022.

———. (2009b). "The Effect of Crystallization of PLA on the Thermal and Mechanical Properties of Microfibrillated Cellulose-Reinforced PLA Composites." *Composites Science and Technology*, 69 (7–8), 1187–92. https://doi.org/10.1016/j.compscitech.2009.02.022.

———. (2010). "Thermo-Mechanical Properties of Microfibrillated Cellulose-Reinforced Partially Crystallized PLA Composites." *Cellulose*, 17 (4), 771–78. https://doi.org/10.1007/s10570-010-9419-5.

Tábi, Tamás, Tatyana Ageyeva, and József Gábor Kovács. (2022). "The Influence of Nucleating Agents, Plasticizers, and Molding Conditions on the Properties of Injection Molded PLA Products." *Materials Today Communications*, 32, https://doi.org/10.1016/j.mtcomm.2022.103936.

Vadori, Ryan, Amar K. Mohanty, and Manju Misra. (2013). "The Effect of Mold Temperature on the Performance of Injection Molded Poly(Lactic Acid)-Based Bioplastic." *Macromolecular Materials and Engineering*, 298 (9), https://doi.org/10.1002/mame.201200274.

Vanheusden, Chris, Pieter Samyn, Bart Goderis, Mouna Hamid, Naveen Reddy, Anitha Ethirajan, Roos Peeters, and Mieke Buntinx. (2021). "Extrusion and Injection Molding of Poly(3-Hydroxybutyrate-Co-3-Hydroxyhexanoate) (Phbhhx): Influence of Processing Conditions on Mechanical Properties and Microstructure." *Polymers*, 13 (22), https://doi.org/10.3390/polym13224012.

Verbeek, Casparus J. R., and Lisa E. van den Berg. (2011). "Development of Proteinous Bioplastics Using Bloodmeal." *Journal of Polymers and the Environment*, 19 (1), https://doi.org/10.1007/s10924-010-0232-x.

Vicente, Diogo, Diogo Neves Proença, and Paula V. Morais. (2023). "The Role of Bacterial Polyhydroalkanoate (PHA) in a Sustainable Future: A Review on the Biological Diversity." *International Journal of Environmental Research and Public Health*. https://doi.org/10.3390/ijerph20042959.

Willberg-Keyriläinen, Pia, Hannes Orelma, and Jarmo Ropponen. (2018). "Injection Molding of Thermoplastic Cellulose Esters and Their Compatibility with Poly(Lactic Acid) and Polyethylene." *Materials*, 11 (12), https://doi.org/10.3390/ma11122358.

Yang, Yu, Jian Min, Ting Xue, Pengcheng Jiang, Xin Liu, Rouming Peng, Jian Wen Huang, et al. (2023). "Complete Bio-Degradation of Poly(Butylene Adipate-Co-

Terephthalate) via Engineered Cutinases." *Nature Communications*, 14 (1), https://doi.org/10.1038/s41467-023-37374-3.

Zee, Maarten Van Der. (2021). "Biodegradability of Biodegradable Mulch Film." *Wageningen Food & Biobased Research*. https://edepot.wur.nl/544211.

Chapter 4

The Effect of Moringa Oleifera Leaves on Thermo-Mechanical Properties of Low-Density Polyethylene

Asim Mushtaq*, PhD,
and Raza Muhammad Khan, ME
[1]Department Polymer and Petrochemical Engineering,
NED University of Engineering & Technology, Karachi, Sindh, Pakistan

Abstract

The increasing influence of utilizing natural fiber composites and products throughout the globe stems from the international environmental conservation group's focus on environmental and ecological issues such as trash recycling, stockpiling, and dumping. Because of the cellulose structure, plant-based fiber may be used as a natural filler in the manufacturing of biocomposites to reduce the negative environmental impact of synthetic –fiber composites. The use of such filler in the creation of composites is critical for reducing the environmental difficulties associated with petro-based reinforced composites. In the present study, composites of Low-density polyethylene (LDPE) have been prepared with Moringa Oleifera Leaves powder as reinforcing filler. This Moringa leaves powder was added to 0 wt. %, 5 wt. % and 10 wt. % in LDPE. The latter aim is to observe this filler's effect on the properties of LDPE by comparing each formulated sample with pure LDPE. The influence of the natural filler Moringa Oleifera leaves powder sourced from plants, on selected mechanical properties of LDPE has been documented in this study. Thermogravimetric analysis (TGA) is used to characterize thermal

* Corresponding Author's Email: engrasimmushtaq@yahoo.com.

In: Injection Molding
Editor: Jose D. Phillips
ISBN: 979-8-89113-429-4

characteristics and Fourier transforms infrared analysis (FTIR) is employed for the functional composition of samples. Moreover, the biodegradability of Moringa- LDPE composites of different formulations is also investigated in this work. Thermo-mechanical properties of the composite were also measured to decide the application of the product.

Keywords: elongation, flexural strength, impact strength, low-density polyethylene, moringa oleifera, tensile strength

Abbreviations

TGA	Thermogravimetric analysis
MOSF	Moringa Oleifera Seed Filler
PEG	Polyethylene Glycol
FMCG	Fast Manufacturing Consumer Goods
PET	Poly Ethylene Terephthalate
FTIR	Fourier Transform Infrared Analysis
HDPE	High-Density Polyethylene.
SDGs	Sustainable Development Goals
LCA	Life-Cycle Assessment
PETG	Polyethylene Terephthalate Glycol
LDPE	Low-Density Polyethylene.
PP	Polypropylene
MO	Moringa Oleifera
r-HDPE	Recycled High-Density Polyethylene.
LDPE	Low-Density Polyethylene
R-PET	Recycled poly (Ethylene Terephthalate)
MFI	Melt Flow Index
LLDPE	Linear Low-Density Polyethylene.

Introduction

Plastic materials like low-density polyethylene, high-density polyethylene, polypropylene, and other common materials have a longer lifespan because of their difficult disposal. These materials can create environmental risks. The usage of biodegradable polymers or composites with natural fillers can help minimize the quantity of plastic waste that ends up in landfills. Natural fiber-reinforced polymer composites have surpassed synthetic fiber composites due to a variety of benefits including cheap cost, lightweight, non-toxic, non-abrasive, and degradable qualities. These are mostly obtained from trees, animals, and minerals. Natural (plant-based) fibers include sisal, coconut, abaca, bamboo, cotton, flax, hemp, banana, jute, and pineapple. Hence in today's world, natural fibers and fillers reinforced with polymer matrix may be found in a variety of semi-structural and lightweight applications in the construction and building industry, as well as the car and aerospace sectors and the packaging industry. "Green composites" are natural fiber-filled polymer composites (Aravindh et al., 2022; Ramesh et al., 2021).

Synthetic fiber composites are not biodegradable, harm the atmosphere by emitting toxic emissions (40 million tons per year), and are expensive. Some examples of synthetic fibers include glass, silica, carbon, and ceramic. Natural fibers, on the other hand, are abundant in nature and do not affect the ecosystem. Natural fiber composites have been shown in studies to be conventional alternatives for synthetic fiber composites. Compared to synthetic fiber processing, natural fiber production has smaller environmental impacts. Figure 1 shows that natural fibers have an increased fiber component with the same efficiency, reducing the base polymer content, which is more polluting. Lightweight bio composites improve energy efficiency and reduce emissions during component use, especially in automotive implementations. End-of-life composting of natural fibers contributes to renewable electricity and carbon allowances. Natural fiber-reinforced composites are not only lightweight, relatively cheap, and abrasion-resistant; they are both non-toxic and operate well (Atmakuri et al., 2020; Ofem et al., 2020).

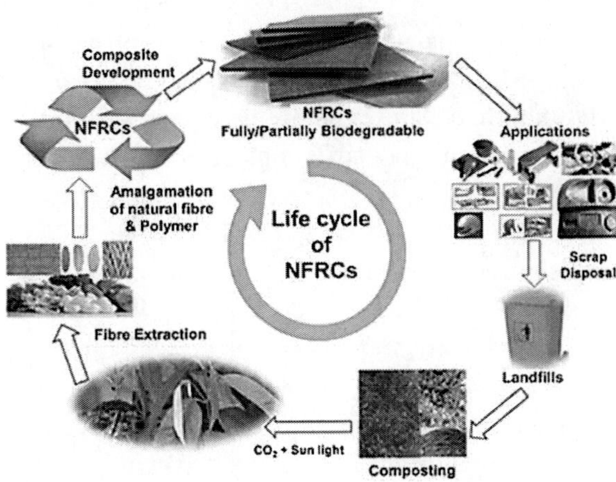

Figure 1. Natural fiber composite life cycle.

Many researchers are now working on natural fibers as synthetic fibers not only hasten global warming but also certain plastic materials are derived from nonrenewable commodities such as oil. The increased use of synthetic fabrics has eventually resulted in environmental issues such as dumping and recycling. Furthermore, synthetic fibers like glass fiber can cause severe skin, eye, and respiratory tract irritation. Long-term illnesses such as cancer and lung scarring have sparked the most concern. On the other hand, when released, glass fiber does not decompose, resulting in environmental waste as well as threats to animal life and wildlife. As a result, one alternative is to use natural fibers rather than synthetic fibers in the development of composite materials because they are recycled. Consumption of natural energy will also provide a good picture of the protection of the green world. The increase in environmental consciousness and societal interest, as well as new environmental regulations and excessive petroleum consumption, caused people to think about using environmentally friendly goods (Al-Tayyar et al., 2020; Arman Alim et al., 2022).

Synthetic fillers are commonly used to change the properties of different polymers but synthetic fillers can cause hazards to the environment. Natural fillers are used in place of synthetic fillers so that the hazard to the environment can be reduced. To reduce the risk to the environment, United Nations Organization has developed different Sustainable Development Goals (SDSs). There are a total of seventeen goals which are known as Sustainable Development Goals (SDGs). The problem stated in this study targets two main goals of SDGs. The existing technique of production and application of plastics is unsustainable. Contamination caused by plastics has been intensified by the increasing manufacturing and use of plastics, particularly single-use plastic goods. Plastic pollution influences endanger animal life and pose a health risk to humans, and ecosystems. There's simply too much plastic to deal with, and recycling isn't sufficient on its own. Plastic pollution is a complex problem that needs a varied approach (Korte et al., 2021; Paidari et al., 2021; Suhag et al., 2020). Natural fillers are used as reinforcing components, and thermoplastic matrix composites are used in a variety of applications. In previous literature, a composite film was made of Moringa Oleifera seed filler (MOSF) and polyvinyl alcohol thermoplastic polymer matrix. The film was prepared by combining PVA with MOSF utilizing a solvent casting approach that included glycerol (plasticizer), glutaraldehyde, and citric acid to create unique biocomposite, ecologically friendly films. MOSF as a possible PVA matrix reinforcement is studied for surface modification, characterization, renewability, and impermeability. When it comes to environmental concerns, biodegradable polymers are an excellent alternative to traditional food packaging (Gangil et al., 2022; Ramesh et al., 2021).

Another study has been done on PBAT (polybutylene adipate-co-terephthalate) and Moringa Oleifera seed film. PBAT (polybutylene adipate-co-terephthalate) is a biocompatible synthetic polymer with a lot of flexibility and a lot of promise for functionalization with bactericidal, antibacterial, and antifungal properties. The casting approach to create PBAT films from Moringa Oleifera (MO) seed powder with varying MO content for use in strawberry food packaging. Morphology, chemical composition, thermal stability, and crystallinity were used to evaluate the MO characteristics. The films were characterized using mechanical tests, and other several application performances. In terms of mechanical and thermal properties, the PBAT films containing 1 percent (wt.) MO was the best for applications such as packaging. MO exhibited no strong associations with the PBAT matrix and lowered the crystallinity of the material, resulting in lower gas and condensate wettability. The PBAT-1 percent MO films worked effectively as biodegradable strawberry packaging, increasing the storage term and reducing the risk of fungal infection. Furthermore, the MO decreased fungal infection when compared to strawberries placed in clean PBAT. The PBAT/MO films suggest that they have a lot of potential for use as active food packaging (Aravindh et al., 2022; Vinod et al., 2021).

The influence of two natural fillers generated from food industry waste products, wheat bran, and pumpkin seeds, on particular mechanical characteristics of low-density polyethylene (LDPE), molded pieces are investigated. The microstructure and fundamental mechanical characteristics of the polymer blends are investigated. In recent years, natural filler polymer blends have become increasingly popular in the polymer processing industry. Polymer mixtures contain a wide range of qualities, which explains why they're employed in so many sectors. Fillers are used to physically modify polymers to get the desired functional qualities. Relationships between the polymer matrix and the filler have a big impact on the final composite material's characteristics. Powdered fillers operate as heterogeneity nucleates in the polymer crystal-like structure in rare situations, especially when their content is low (Atmakuri et al., 2020; Bhardwaj et al., 2022).

The purpose of the study was to govern how powdered fillers made from renewable resources like pumpkin seed hulls and wheat bran influenced the mechanical properties of low-density polyethylene filled with them. The interactions amid the fundamental strength attributes of injection molded components and the filler content in particular were investigated in this study. It also used a matrix-filler system analysis to try to explain the observed behaviors. Because of its cellulose nature, agricultural waste may be used as a natural filler in the manufacture of green composites to reduce the negative impact on the environment. Due to their outstanding mechanical qualities, cheap cost, and low density, agricultural waste-based green composites can take the place of hydrocarbon fiber composites. Recent breakthroughs in the field of natural fibers, particularly green composites made from agricultural waste fibers (Dehury et al., 2021; Ofem et al., 2020). Three distinct thermosets of electrically charged protective coating wastes were employed as fillers for low-density polyethylene. Composites of LDPE were prepared with some organic filler (polyester fiber) and inorganic fillers (glass fiber, oil shale ash, limestone ash, mineral wool fiber) in the form of waste because they cause massive land fillings and pollute the environment. The second purpose is to look into the effects of organic and inorganic fillers on the mechanical properties of LDPE to improve the flexural and tensile properties of the material. The following stage is to add fillers such as mineral wool fiber, oil shale ash, and limestone ash to increase the heat conductivity and cooling rate of the LDPE composite to increase the processing speed for large-scale production (Ashok and Kani, 2021; Bhardwaj et al., 2022).

The main objective of that study was to figure out how fillers affect LDPE's tensile and flexural properties, as well as the effect of compatibilizers on mechanical qualities, and to employ fillers to improve LDPE's flexural capabilities. Plastic production has eclipsed steel production in volume during the last five years. Plastics are widely used in vehicle components, aircraft parts, and consumer items because of their lightweight, simple cycle capacity, and corrosion resistance. Inorganic fillers have been replaced with normal fillers in polymer composites due to environmental awareness, concerns, and new requirements. The main reasons for using regular fillers widely in the development of polymer composites were their abundance, sustainability, low cost, non-harmfulness, low thickness, and reduced abrasiveness during handling. "Green composites" refers to normal fiber-filled polymer composites (Al-Tayyar et al., 2020; Gulihonenahali Rajkumar et al., 2021).

Healthy oceans and seas are essential for human life. They cover 70% of the earth's surface and afford us feed, water, and power. Despite this caused havoc on these precious resources. They must instantaneously begin to manage and defend all marine life on the globe sustainably and one way of doing that is by reducing pollution. This can be done by fabricating a composite by introducing natural filler (Moringa leaves powder) in place of synthetic filler so that the use of the type of composite which can partially be decomposed is possible and the risk to the environment can be lowered. This research aims to synthesize the particle reinforce polymer composite by using a natural filler which is Moringa Oleifera leaves for commodity application (Kenawy and Khalil, 2021; Munteanu and Vasile, 2019; Polmann et al., 2022).

Experiment

Materials

There are two major components in a composite; reinforcing material and matrix. Polymers, ceramics, and metals can all be used as matrix materials. Matrix materials are chosen based on their final purpose. Reinforcing material can take the shape of particles, fiber, sheets, or fabric. In the present research, LDPE is being used as a matrix whereas Moringa Oleifera leaves powder is being used as a reinforcing material. Figure 2 shows the fiber-reinforced composite process flow diagram.

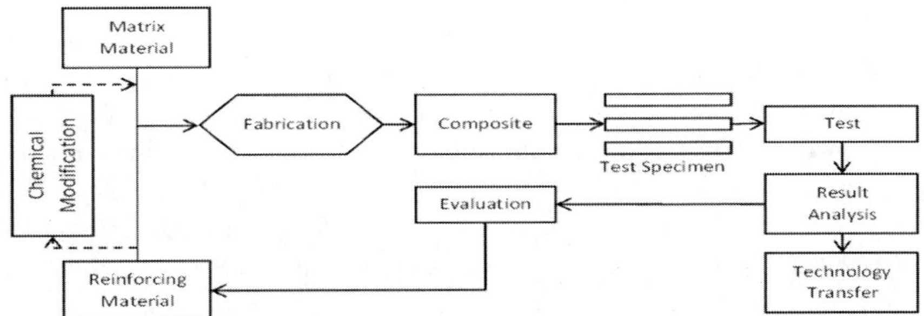

Figure 2. Fiber-reinforced composite process flow diagram.

The Lotrene Mg 70 low-density polyethylene (LDPE) is used and the technical datasheet for this grade is shown in Table 1. The natural filler Moringa Oleifera (M Oleifera) leaves powder has a high potential to be used as filler. These leaves are sourced from Moringa Trees. This plant belongs to the Moringaceae family and is endemic to the sub-Himalayan regions of northwestern India, Pakistan, Bangladesh, and Afghanistan. It is regarded as the most useful plant in the world, because other than its components used as edibles; every part of this tree has immense medical and industrial uses. MOL has received much interest because of its low cost, provides ease of usage, and rapid expansion on a variable scale of location, space, and geographical state. The Moringa leaves used for this study are bought from an organic herbal company (Jena et al., 2021; Sani et al., 2021). The density of Moringa Oleifera leaves is 1.06 ± 0.24 g/cm^3.

Table 1. Properties of LDPE grade

Properties	Value	Unit	Test Method
Elongation @ Break	240	%	ASTM D-638
Crystalline melting point	102	°C	ASTM E-794
Density @ 23°C	0.918	g/cm^3	ASTM D1505
Vicat softening point	80	°C	ASTM D-1525
Tensile strength @ Break	7	MPa	ASTM D-638
Tensile strength @ Yield	8	MPa	ASTM D-638
Melt flow index	70	g/10 min	ASTM D-1238

Sample Preparation

The sample preparation involves; the drying of Moringa leaves powder, formulation and mixing, and fabrication of composites. If Moringa leaves were obtained from the local market then the leaves should be thoroughly cleaned with distilled water to eliminate surface contaminants. These are then followed by sun-drying for 24 hours before being placed in an oven. The leaves are then chopped and separated into bits before being made into powder and ground. Moringa leaves powder is dried just before making each composition in a forced convection oven. At 50 degree Celsius for 1.5 hr. (until the weight of leaves is constant) to assure total evaporation of moisture. The composition ratio of Moringa Oleifera powder in LDPE is; M0 is pure LDPE, M4 is LDPE 95 wt. % with Moringa 5 wt. % and M8 are LDPE 90 wt. % with Moringa 10 wt. %. To make sure the homogeneous mixing of material both the matrix and reinforcement are taken in powder form. Initially, both material is mechanically mixed in a beaker with a high-speed stirrer. Homogeneous mixing of matrix and reinforcement can be assured by this physical mixing. Then the final samples are ready to evaluate for mechanical and thermal characteristics to optimize the fiber content. The equipment used for the process was Laboratory Press (compression molding machine) provided by Gibitre instruments from Italy. The laboratory press is a crucial instrument for processing samples with consistent properties. The basic factors to achieve this result are uniform temperature transmission over the steel plate surface, mechanical stiffness to assure constant sample thickness, and persistent closure force. A hot laboratory press is used for making final samples. The equipment used for the process was Laboratory Press (compression molding machine) provided by Gilberte Instruments from Italy. To mold the material (sheet) into a final and finished shape the mold's cavity is filled with the mixture of LDPE and Moringa powder, mold is closed by applying a pressure of 160 psi (approx.) is applied at the temperature of 125°C approximately. The mold is removed after 3 minutes (approx.) and then cooled. The finished product is then obtained shown in Figure 3.

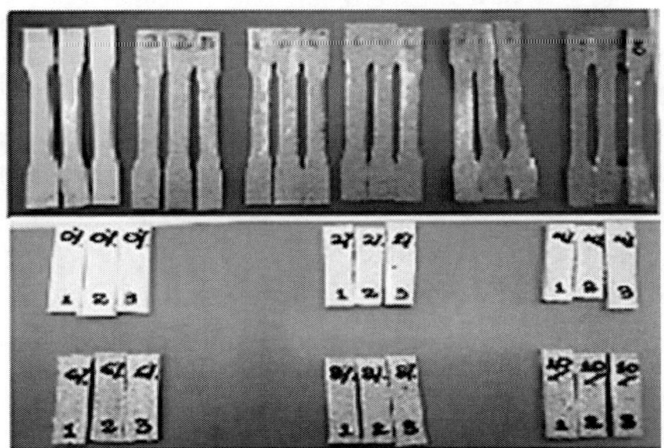

Figure 3. Fabricated Moringa leaves filled composite with LDPE.

Tensile Test

The UTM universal testing machine is used to assess a material's tensile, compression, and flexural strength. The machine used ASTM D638 for tensile test manufacturer by Zwick/Roell (Z005) materials testing machine 5 kN ProLine Model (Z005) and test load 5 kN – 100 kN. The technique should be carried out at room temperature with a certain constant speed. The speed was set at 50 mm/min. The gauge length of the dumbbell-shaped sample was 50 mm. The thickness of the sample was measured to be 3 mm and the width was 12.5 mm. The tensile strength of the samples can be calculated by using Eq. (1) (Jena et al., 2021; Kenawy and Khalil, 2021).

$$\sigma = F/A \tag{1}$$

where; σ represents tensile strength (MPa or N/mm^2), F is the applied load (N), and A is the original area (mm^2) of the sample.

Flexural Test

The UTM machine used for flexural strength measurements of samples is by IDTNT Model number DTU900 – MHA and capacity of 300 kN. Flexural strength can be defined as the stress at failure in bending. ASTM D790 for Universal Testing Machine (UTM) of flexural strength was used to govern the mechanical behavior of the given samples by applying three-point bending stress on the sample material. The speed was set at 10 mm/min. The gauge length of the rectangular-shaped sample was 50 mm. The thickness of the sample was measured to be 3.28 mm and the width was 12.7 mm. For a specimen under a load in a three-point bending setup. The flexural strength can be measured by Eq. (2) (Ashok and Kani, 2021; Gulihonenahali Rajkumar et al., 2021).

$$\sigma = \frac{3FL}{2bd^2} \tag{2}$$

where; L is the length of the support span (mm), F is the load (force) at the fracture point (N), d is the thickness (mm), and b is the width (mm).

Izod Impact Test

Universal Pendulum Impact Tester Version 1.08 by RAY-RAN is employed for Izod test results. The energy absorption capacity of the polymer composites can be determined by the Izod impact test as per ASTM standards. ASTM D256-10 standard for the determination of the impact strength of materials is the Izod impact strength test. A swinging arm is lifted to a predetermined height before being released (constant potential energy). The hammer used was with a velocity of 3.46 m/s and the energy of the hammer was 4 Joules. The arm swoops down and crushes the sample, shattering it. The stature at which the arm swings after hitting the item does not settle the sample's energy absorption and impact. The thicknesses of the samples were measured to be 2.6 mm and the width was 12.2 mm approx.

Thermogravimetric Analysis (TGA)

The TGA analyzer equipment is manufactured by METTLER TOLEDO (Switzerland), Model TGA/DSC 3+. The TGA/DSC 3+ system analyses both heat flow and sample weight fluctuation concurrently using one of many replacement sensors. Thermal effects in the range of up to 1600 °C are measured using a TGA system with a high-temperature oven. The weight measurement on these scales is independent of the sample location, and the internal mass calibration is done automatically. TGA is an experimental technique for determining a material's thermal properties and volatile constituent proportion by weighing a sample while it is heated at a steady rate. Therefore the TGA of unreinforced LDPE and then with Moringa leaves powder composite samples will be carried out with the assistance of During this procedure, a constant heating rate should be maintained through a range of different temperatures and the findings will then be used for thermal analysis of composite samples. The weight of the sample used was between 7 to 13 mg. The samples were heated from 25ºC to 600ºC at a heating rate of 10ºC/min. thermal analysis was performed under a nitrogen environment at a constant flow rate of nitrogen. The data of weight loss and amount of heat flow was obtained during the experiment and was noted and percentage loss in weight was plotted as a function of temperature.

Fourier Transform Infrared (FTIR)

Fourier Transform Infrared analyzer spectrometer can be used for chemical analyses of several materials by collecting their spectrum. The configurations manufacturer is Thermo Fisher Scientific and Model Nicolet iS20 FTIR Spectrometer. The data is collected in the near-IR, mid-IR, and far-IR spectral ranges. The equipment is provided with several equipment accessories to be used according to experiment requirements. The system is integrated with cutting-edge instrument features, including step-scan and dual-channel data collection. The system is accessible in two configurations: single and advanced automated multi-range. This model works in the spectral range of Mid-IR with a spectral resolution of 0.25 cm^{-1} the scan modes from this model are in linear forms.

Biodegradation Analysis of Composites

In a soiled setting, pure LDPE (M0), M4 (LDPE 95 wt. % with Moringa 5 wt. %), and M8 (LDPE 90 wt. % with Moringa 10 wt. %) filled composites were tested for biodegradability. The botanical garden soil was poured into a plastic pot. The specimens under investigation were weighted (W_i) and then buried in the pot. For 16 days and then 36 days, the samples were carefully taken from the soil, cleaned, and reweighed (W_f).

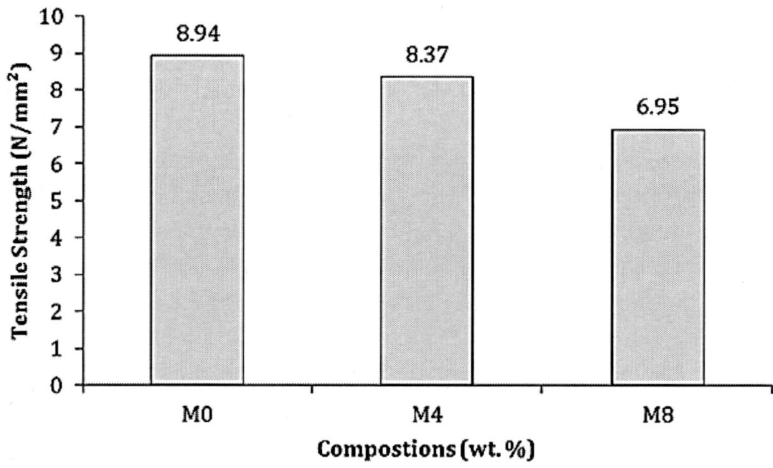

Figure 4. Effect of filler wt. % on tensile strength of LDPE.

Discussion

The three sample composition ratio of Moringa in LDPE are; M0 is pure LDPE, M4 are LDPE 95 wt. % with Moringa 5 wt. % and M8 are LDPE 90 wt. % with Moringa 10 wt. % were fabricated and characterized.

Effect of Different Compositions of Filler on Tensile Strength of LDPE

From Figure 4 it was observed that with increasing the amount of filler the tensile properties decrease. A tensile test involves the application of forces in a uniaxial direction. When tensile stress is applied to the specimen the material starts to elongate in the machine direction. The fall in tensile strength of the composite might be due to poor interaction between the matrix and reinforcement. As the moringa was used in powder form, the small gap between the interface of the matrix and filler may cause a big loss of tensile strength (Gangil et al., 2022; Ramesh et al., 2021; Vinod et al., 2021).

Effect of Different Compositions of Filler on Percent Elongation of LDPE

Figure 5 shows that; between different compositions of LDPE and filler, it can be evaluated that the percent elongation of the sample with compositions LDPE 95 wt. % and Moringa 5 wt. % (M4) is higher than the pure LDPE (M0) sample. Whereas the percent elongation of the sample with compositions LDPE 90 wt. % and Moringa 10 wt. % (M8) is lower as compared to pure LDPE. This result shows that moringa powder encountered some flexibility when used in lower quantities. Increasing moringa quantity may create saturation issues and hence can reduce the properties of a composite (Aravindh et al., 2022; Atmakuri et al., 2020; Dehury et al., 2021).

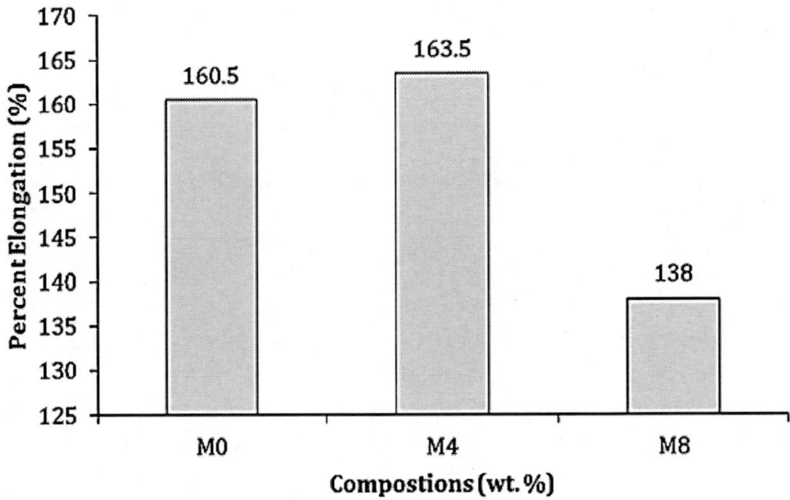

Figure 5. Effect of filler wt. % on percent elongation of LDPE.

Effect of Different Compositions of Filler on Flexural Strength of LDPE

Figure 6 displays that the flexural strength of LDPE decreases as the amount of filler is increasing. The bonding between the matrix and reinforcing filler may be not good enough on a micro level, hence reducing the properties of the sample by weakening intermolecular forces between the constituents. The use of coupling agents or pre-treatment of matrix material can enhance the adhesion between two phases of the composite hence improving the mechanical properties of the composite. Also, the elongation results indicate the energy absorption in the sample, making it more flexible and decreasing the stiffness of the sample (Ashok and Kani, 2021; Gulihonenahali Rajkumar et al., 2021).

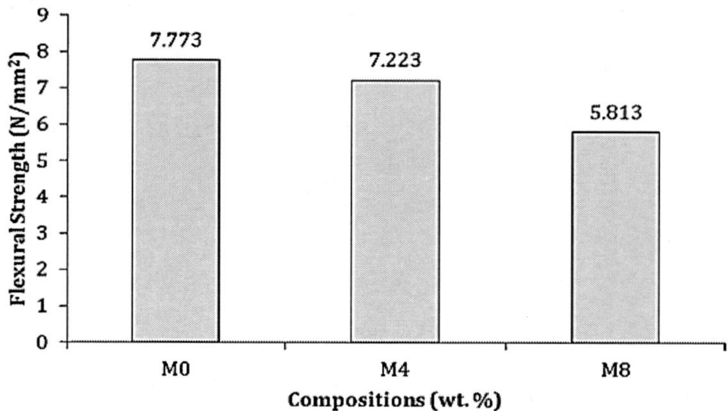

Figure 6. Effect of filler wt. % on flexural strength of LDPE.

Figure 7 represents that the composite with a 5/95 ratio of Moringa and LDPE respectively, absorbed more load before breaking as compared to pure LDPE, whereas as increase the amount of Moringa to 10 wt. % the breaking point load decreases. The breaking point load depends upon the morphology of the composite. The load may shift from the matrix to filler and shows good properties against break point load (Gulihonenahali Rajkumar et al., 2021; Jena et al., 2021).

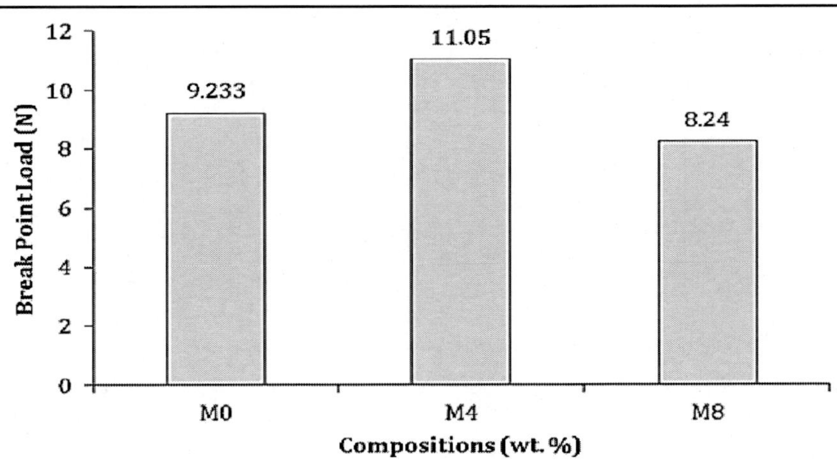

Figure 7. Effect of filler wt. % on break point load of LDPE.

Effect of Different Compositions of Filler on Impact Strength of LDPE

Figures 8 and 9 show the energy absorbed and impact strength without a notched sample by the sample with the composition ratio of 5 wt. % Moringa is greater as compared to pure LDPE. Hence the impact strength with 5 wt. % Moringa is greater as compared to pure LDPE sample but as the amount of Moringa is increased to 10 wt. % of the energy absorbed without a notched sample decreases resulting in decreased impact strength. There might be variation in the results because all the samples were tested under un-notched conditions. With the same testing condition, moringa with a ratio of 5% will absorb more energy than pure low-density polyethylene.

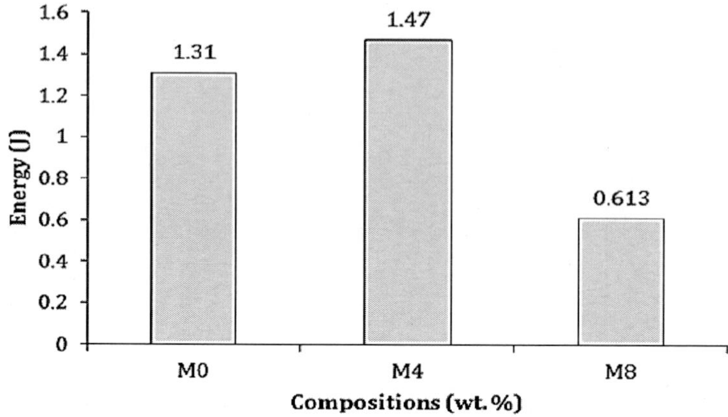

Figure 8. Effect of filler wt. % on energy absorbed without notch sample of LDPE.

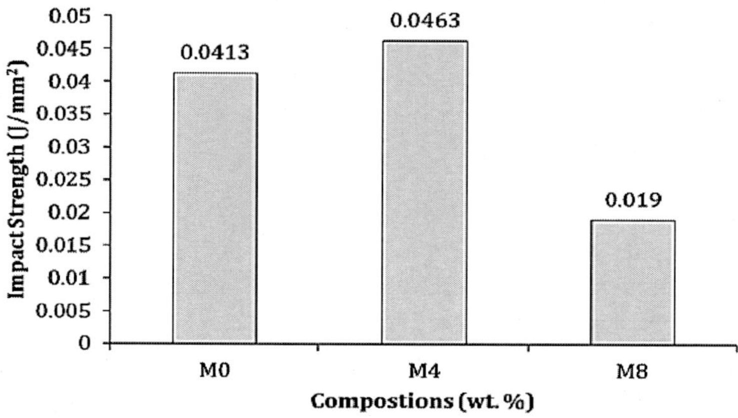

Figure 9. Effect of filler wt. % on the impact strength of LDPE.

Effect of Different Compositions of Filler on Thermal Gravimetric Analysis of LDPE

Figure 10 shows that the TGA was performed for the pure LDPE, LDPE 95 wt. % with 5 wt. % Moringa and LDPE 90 wt. % with 10 wt. % Moringa, in which the weight loss due to volatilization of the degraded polymer chain was observed as a function of temperature. It is observed that the visible thermal degradation of pure LDPE started at 380.66°C and ended at 495°C. For the sample with 5 wt. % Moringa content the degradation ranges from 380°C to

500ºC. For the sample with 10 wt. % Moringa leaves the first degradation was at 348ºC which is due to the evaporation of volatile constituents of Moringa. The degradation of polymer chains then started at 380.84ºC and ended at 500ºC. It was also observed that the onset degradation of LDPE-Moringa composites was almost the same as the pure LDPE when the composition ratio was 5/95, but there is a very slight change in the onset temperature of the sample with a composition ratio of 10/90. The early degradation can be due to the presence of moisture or volatile constituents. It can also be observed that for all samples, once the weight loss was started, it occurred at an accelerated rate, drastically reducing within the narrow range of temperature. This was due to a nucleation-derived mechanism, according to which the process starts at discrete points called nuclei and from where the degradation spreads into the rest of the composite. Overall, it can be seen that the Moringa leaves reinforcement into the LDPE is not disturbing the thermal properties of pure LDPE and more or fewer composites of different compositions are showing the same results; hence thermal stability of LDPE remains intact and unaffected (Gulihonenahali Rajkumar et al., 2021; Jena et al., 2021).

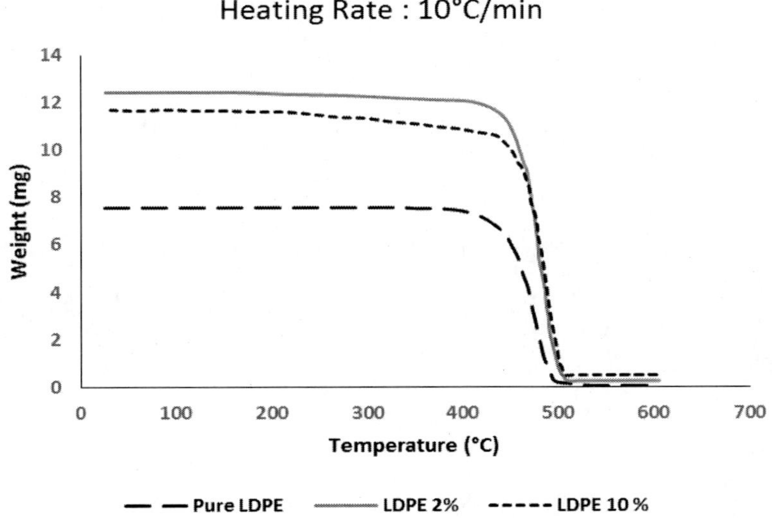

Figure 10. Effect of filler wt. % on thermal gravimetric analysis of LDPE.

Effect of Different Compositions of Filler on FTIR of LDPE

Figure 11 shows that there is a major peak at 1597 cm^{-1} shown in FTIR of pure Moringa leaves. This peak is the main differentiator which shows the presence of cellulose content in Moringa leaves whereas in pure LDPE there is no such peak shown in Figure 12. The magnified cellulose peak is present in FTIR of both the composites of composition M4 (5 wt. % Moringa 95 wt. % LDPE)

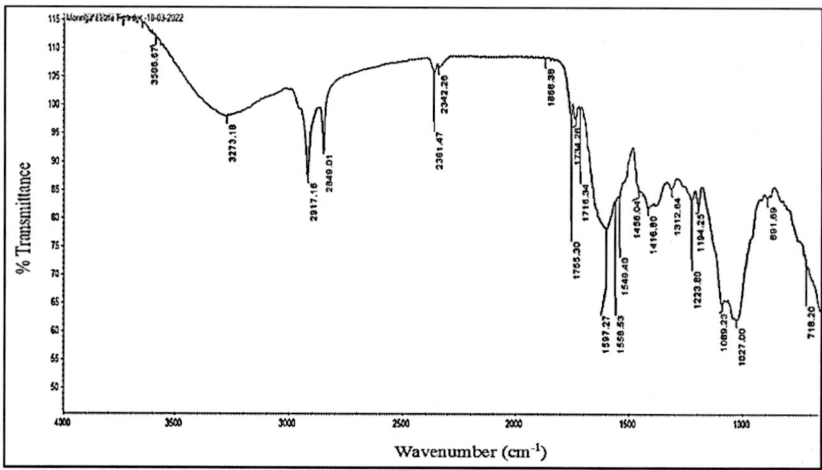

Figure 11. FTIR of pure Moringa.

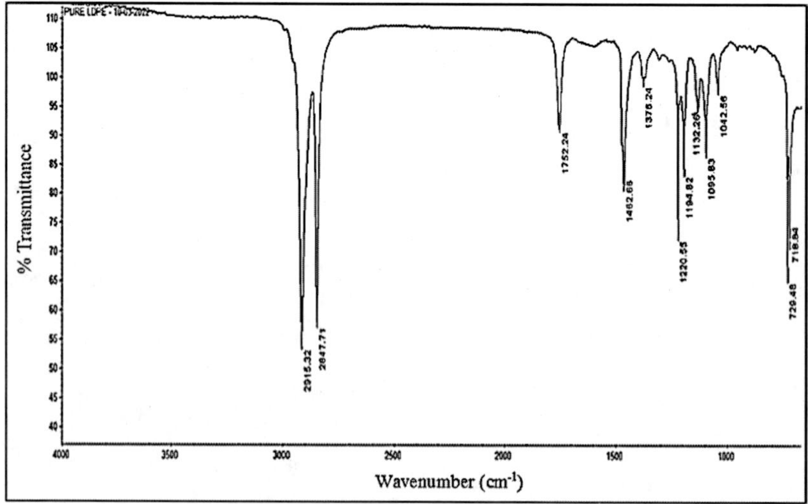

Figure 12. FTIR of pure LDPE.

and M8 (10 wt. % Moringa 90 wt. % LDPE) is shown in Figure 13. This peak appeared at 1596 cm^{-1} assigned for C = O stretching for cellulose contents and hence the FTIR analysis the presence of cellulose is confirmed in the composites (Aravindh et al., 2022; Dehury et al., 2021).

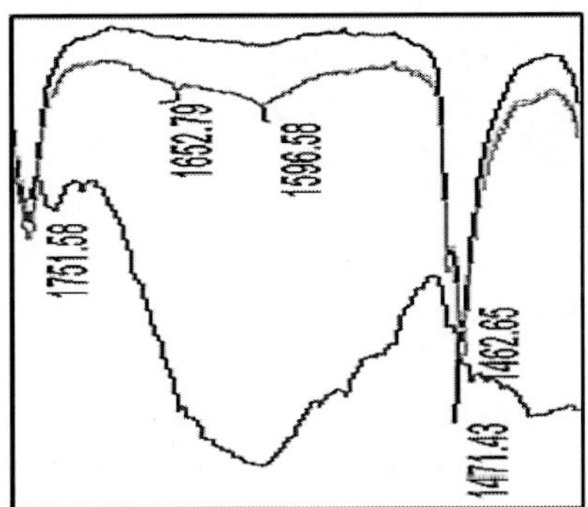

Figure 13. Magnified cellulose peak composition M4 (5 wt. % Moringa 95 wt. % LDPE) and M8 (10 wt. % Moringa 90 wt. % LDPE) compare pure LPDE.

Effect of Different Compositions of Filler on Biodegradability of LDPE

Table 2 shows that there is no degradation or weight loss examined in samples as the weight remained constant. It is proved from the studies, that once the weight loss started, it occurred at an accelerated rate, drastically reducing the weight due to a nucleation-derived mechanism, according to which the process starts at discrete points called nuclei and from where the degradation spreads into the rest of the composite. It can be evaluated that still that threshold/discrete time isn't reached yet to start the nucleation-derived mechanism for degradation. Composites need to be placed under the soil for some more time (Arman Alim et al., 2022; Ramesh et al., 2021; Sani et al., 2021).

Table 2. Effect of filler wt. % on biodegradability of LDPE

Samples	Initial weight (W_i) gm	Final weight (W_f) gm	
		After 14 days	After 34 days
M0	6.75	6.75	6.75
M4	6.61	6.61	6.61
M8	6.61	6.61	6.61

Conclusion

In this research, the effect of moringa powder was analyzed on the thermal and mechanical properties of low-density polyethylene. Moringa plant leaves under normal room conditions. The dried leaves are then crushed and converted into powder. LDPE is also pulverized to attain the maximum surface interaction between matrix and filler. Moringa powder absorbs moisture from the air. So before mixing it with LDEP the powder was dried in the oven and then mixed with LDPE power with the help of a mechanical stirrer. The physically mixed material was then put into a hot press. At this melt processing stage, a composite filler and matrix were formed. Different sample for mechanical testing was manufactured by using a hot press. The melt mixed material was now used for thermal and mechanical characterization. From the results of mechanical tests, it is perceived that the tensile strength and flexural strength both decrease with the incorporation of Moringa Oleifera leaves powder as filler in the LDPE matrix. But on the other hand percent elongation as well as impact strength of composite with 5 wt. % Moringa as filler shows enhanced results. The enhanced impact strength indicates the better performance of the material under impact load without fracture. Through the overall TGA results, it can be seen that the Moringa leaves reinforcement into the LDPE is not disturbing the thermal properties of pure LDPE and more or fewer composites of different compositions are showing the same results; hence thermal stability of LDPE remains to interact and unaffected. For the functional composition of samples, FTIR was employed. FTIR showed the presence of cellulose content in samples of different formulations hence proving the incorporation of Moringa Leaves filler into the LDPE. There is no degradation or weight loss examined in samples as still the discrete-time to start degradation isn't reached yet. Composites need to be placed under the soil for some more time to reach that discrete point to start degradation. In the future study, it is recommended to characterize the mechanical and thermal properties of these composites after employing mixing techniques, such as mechanical mixing by a twin screw extruder. The molding technique can also be changed and injection molding technique can be used. The pretreatment of the LDPE matrix can also be done to enhance bonding between the matrix and reinforcement molecules. Moreover, a coupling agent can be added to exclude phase separation between the two constituents. The early degradation can also be achieved by introducing moisture content on daily basis to increase microbial effects in the composites. According to the results, the composite can be used in any application where good energy absorption property is required in the product.

References

Al-Tayyar N. A., Youssef A. MAl-hindi R. Antimicrobial food packaging based on sustainable Bio-based materials for reducing foodborne Pathogens: A review. *Food Chem* (2020) 310(1):1-17.

Aravindh M., Sathish S., Ranga Raj R., Alagar Karthick, V. Mohanavel, Pravin P. Patil, M. Muhibbullah and Sameh M. Osman. A Review on the Effect of Various Chemical Treatments on the Mechanical Properties of Renewable Fiber-Reinforced Composites. *Adv Mater Sci Eng* (2022) 2022(1):1-24.

Arman Alim A. A., Mohammad Shirajuddin S. S., Anuar F. H., A Review of Nonbiodegradable and Biodegradable Composites for Food Packaging Application. *J Chem* (2022) 2022(1):1-27.

Ashok K. G., Kani K. Experimental studies on interlaminar shear strength and dynamic mechanical analysis of luffa fiber epoxy composites with nano PbO addition. *J Ind Text* (2021) 51(3):3829-3854.

Atmakuri A., Palevicius A., Vilkauskas A., Giedrius Janusas. Review of Hybrid Fiber Based Composites with Nano Particles-Material Properties and Applications. *Polymers* (2020) 12(9):1-30.

Bhardwaj S., Lata S. Garg R. Application of nanotechnology for preventing postharvest losses of agriproducts. *J Hortic Sci Biotechnol* (2022) 97(1):1-14.

Dehury J., Mohanty J. R., Nayak S., PriyaRanjan Samal, Sujit Kumar Khuntia, Chandrabhanu Malla, Saumya Darsan Mohanty & Jagannath Mohapatra. Comprehensive Characterization of Date Palm Petiole Fiber Reinforced Epoxy Composites: Effect of Fiber Treatment and Loading on Various Properties. *J Nat Fibers* (2021) 18(1):1-14.

Gangil B., Ranakoti L., Verma S. K., Tej Singh. Utilization of waste dolomite dust in carbon fiber reinforced vinylester composites. *J Mater Res Technol* (2022) 18(1):3291-3301.

Gulihonenahali Rajkumar A., Hemath M., Kurki Nagaraja B., Shivakumar Neerakallu, Senthil Muthu Kumar Thiagamani, Mochamad Asrofi. An artificial neural network prediction on physical, mechanical, and thermal characteristics of giant reed fiber reinforced polyethylene terephthalate composite. *J Ind Text* (2021) 51(1):769-803.

Jena P. K., Samal P., Nayak S., Jyoti Ranjan Behera, Sujit Kumar Khuntia, Jagannath Mohapatra, Saumya Darsan Mohanty & Chandrabhanu Malla. Experimental Investigation on the Mechanical, Thermal, and Morphological Behaviour of Prosopis juliflora Bark Reinforced Epoxy Polymer Composite. *J Nat Fibers* (2021) 18(1):1-11.

Kenawy S. H. Khalil A. M. Reclaiming Waste Rubber for a Green Environment. *Biointerface Res Appl Chem* (2021) 11(1):8413-84423.

Korte I., Kreyenschmidt J., Wensing J., Stefanie Bröring, Jan Niklas Frase, Ralf Pude, Christopher Konow, Thomas Havelt, Jessica Rumpf, Michaela Schmitz and Margit

Schulze. Can Sustainable Packaging Help to Reduce Food Waste? A Status Quo Focusing Plant-Derived Polymers and Additives. *Appl Sci* (2021) 11(11):1-41.

Munteanu S. B. Vasile C. Vegetable Additives in Food Packaging Polymeric Materials. *Polymers* (2019) 12(1):1-34.

Ofem M. I., Ene E. B., Ubi P. A., S. O. Odey, D. O. Fakorede. Properties of cellulose reinforced composites: A review. *Nigerian J Tech* (2020) 39(2):386-402.

Paidari S., Zamindar N., Tahergorabi R., Maryam Kargar, Shima Ezzati, Nadia shirani & Sayyed Hossein Musavi. Edible coating and films as promising packaging: a mini review. *J Food Meas Charact* (2021) 15(5):4205-4214.

Polmann G., Badia V., Danielski R., Sandra Regina Salvador Ferreira & Jane Mara Block. Nuts and Nut-Based Products: A Meta-Analysis from Intake Health Benefits and Functional Characteristics from Recovered Constituents. *Food Rev Int* (2022) 38(1):1-27.

Ramesh M., Rajeshkumar L., Bhuvaneswari V. Leaf fibres as reinforcements in green composites: a review on processing, properties and applications. *Emergent Mater* (2021) 5(3):833-857.

Sani M. A., Azizi-Lalabadi M., Tavassoli M., Keyhan Mohammadi, David Julian McClements. Recent Advances in the Development of Smart and Active Biodegradable Packaging Materials. *Nanomaterials* (2021) 11(5):1-34.

Suhag R., Kumar N., Petkoska A. T., Ashutosh Upadhyay. Film formation and deposition methods of edible coating on food products: A review. *Food Res Int* (2020) 136(1):1-16.

Vinod A., Sanjay M. R., Siengchin S., Steffen Fischer, Fully bio-based agro-waste soy stem fiber reinforced bio-epoxy composites for lightweight structural applications: Influence of surface modification techniques. *Constr Build Mater* (2021) 303(1):1-17.

Chapter 5

Usability of Natural Fibers in Thermoplastic Composite Materials

Ilyas Kartal[1,*], PhD
and Hilal Selimoglu[2], Msc
[1]Department of Metallurgical and Materials Engineering, Marmara University, Istanbul, Turkey
[2]Department of Metallurgical and Materials Engineering, Institute of Pure and Applied Sciences, Marmara University, Istanbul, Turkey

Abstract

Thermoplastic-based composites consist of a thermoplastic matrix and reinforcement materials. Reinforcement material is used to strengthen the matrix and improve its properties. Both fiber use and filler use are common in composites. Inorganic fibers consist of non-organic materials such as metal, ceramic, and glass and are extensively used in the production of thermoplastic-based composite materials. However, the use of natural fibers instead of inorganic fibers has recently become widespread. Natural fibers consist of materials that are of plant, animal, or mineral origin. Natural fibers are widely used in the production of composite materials due to their advantages, such as low cost, good environmental performance, biodegradability, and sustainability. Composite materials can be made extremely strong and stiff by adding natural fibers like flax, bamboo, and wood fiber. Natural particles such as sawdust can be used to impart mechanical, thermal, and chemical properties to composite materials. Fillers can also be used to increase the volume and weight of composite materials. However, due to the structure of natural fibers, they react with water. This causes weak bonds at the

* Corresponding Author's Email: ilyaskartal@marmara.edu.tr.

In: Injection Molding
Editor: Jose D. Phillips
ISBN: 979-8-89113-429-4

composite interface and negatively affects the properties of the composite. In composites, interface enhancement takes place for natural fibers. Thus, impurities are eliminated, fiber-matrix interface bonding is improved, and the performance of the composite is increased. Natural filler-filled composite materials are used in many different fields, such as automotive, construction, aerospace, defense, packaging, and agriculture.

This study explains issues such as the advantages and usability of natural fiber reinforcement polymer composites in terms of environmental sustainability. Factors affecting the properties of natural fiber-reinforced composites, interface improvements, degradation, use of recycled polymers with natural fibers, and production methods are mentioned. Additionally, current usage areas are also explained. This research provides important guidance to researchers and industry professionals who want to promote the use of natural fillers and fibers in thermoplastic-based polymer composites and contribute to sustainable material solutions in the future.

Keywords: natural fiber reinforcement composite, thermoplastic, surface treatments, natural fillers

Introduction

In recent years, the development of materials with good mechanical properties has become a priority issue among researchers. As a result of this long-standing search, composite materials consisting of two or more components have been developed instead of single-phase materials. Composite materials are one of the most beautiful discoveries and the fastest-growing field of contemporary science. By changing the chemical and physical properties of the components, the properties of the final composite can be optimized. Composites can be found in every material class, from metals to ceramics to polymers. Among these, thermoplastic and thermoset polymers have come to the fore in the use of matrix materials, partly due to their low cost and high durability [1].

Composite materials consist of at least two materials: matrix and reinforcement/filler. When one or both of the reinforcements or matrix components of this group are derived from plant sources, the composite is referred to as green composite. For green composites, terms like bio, renewable, and sustainable are employed. These expressions are used for materials produced from sustainably sourced plant-based raw materials found in nature [2–3].

Green composites are divided into three main categories according to reinforcement and polymer material types.

(a) Fully renewable composites in which both matrix and reinforcement are from renewable sources;
(b) Partially renewable composites in which the matrix is obtained from renewable resources and reinforced with a synthetic material;
(c) They are partially renewable composites in which the synthetic matrix is reinforced with renewable resources.

According to the sources of natural fibers used in green composites, bast fibers (jute, kenaf, flax, etc.), leaf fibers (palm, sisal, manila, etc.), seed fibers (cotton, kapok, etc.), fruit fibers (coconut, coco fiber, etc.), stem fibers (bamboo, rice, grass, corn, wheat, etc.), and wood-based fibers (chestnut wood fiber, pine wood fiber, etc.) [4-6]

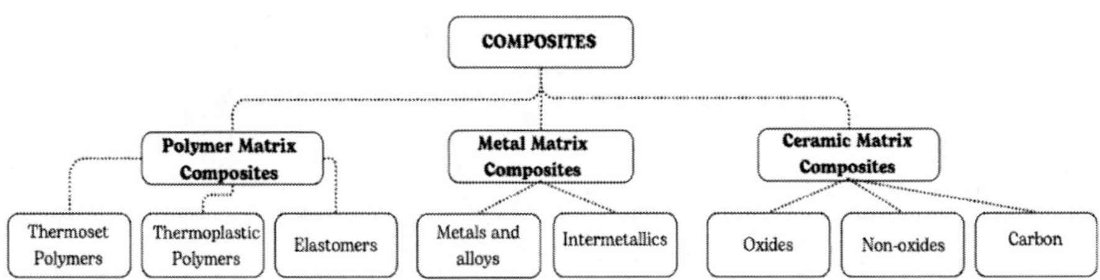

Figure 1. Classification of composite materials according to matrix material.

Composites created using synthetic fibers are expensive, heavy, and difficult to produce, and they have negative effects on the environment and human health. The production of synthetic fibers is 10 times more energy-intensive than the production of natural fibers and is largely dependent on fossil fuels. As a result, synthetic fiber production emits more CO_2 into the environment than natural fiber production. For example, polypropylene fiber production consumes 20 times more energy than natural fibers and has negative effects such as global warming, and depletion of fossil resources [7].

Natural fiber-reinforced polymer composites (NFC) are unique materials for a wide variety of applications, including aircraft, automobiles, construction, and trains. Due to their advantages such as low price, renewability, biodegradability, recyclability, and low-density properties, natural fibers, and fillers are used as potential reinforcement materials in polymer matrices instead of many synthetic fibers such as glass [8].

Plastic waste is a concern for environmental health management due to its impact on the environment. Reducing plastic waste through treatment methods like recycling, reuse, energy recovery, and landfilling has become a movement to address this issue and lessen pollution in the environment. Efforts continue to make recycling waste a global necessity to produce value-added materials that can be a social solution. They can compete with synthetic or inorganic fillers because they are less dependent on fossil sources. Filler and fiber production from renewable raw materials is expected to increase from 12% in 2010 to approximately 25% in 2030 [9].

Although the use of natural fibers or fillers is widely preferred in thermoplastics, their use is also common in thermosets. Many studies have been carried out in the literature on natural fiber-reinforced thermoset composites [10–12].

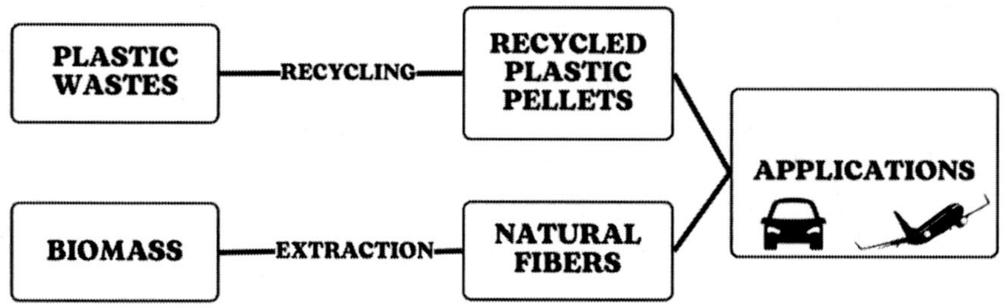

Figure 2. Schematic representation of plastic waste recycling and natural fiber-based composite production.

Factors Affecting the Properties of Natural Fiber-Reinforced Composites

One of the most important features that affect the mechanical properties of composite materials is the fiber-matrix interface properties. The interface that acts as a bond between the fiber and the matrix can be a physical, chemical, or mechanical bond, depending on the type of fiber and matrix materials. When a load is applied to a composite, a stress transfer must occur between the reinforcement and the matrix for the entire structure to carry the load. The interface must transfer the load applied to the fiber to the matrix. As a result, the applied load can be supported throughout the whole composite. It is possible to determine many mechanical behaviors of composite materials by accurately determining the stress transfer mechanism and stresses between the fiber and matrix.

Composite materials with a strong interface have high strength and low ductility. Materials with a weak interface structure have low strength and high fracture resistance. The mechanical properties of composite materials formed by combining different fiber and matrix materials can be calculated accurately by determining the interface properties. Mechanical testing conducted on other materials cannot be applied to the interface due to its thin-layer nature. For this reason, various methods have been developed to determine the interface properties.

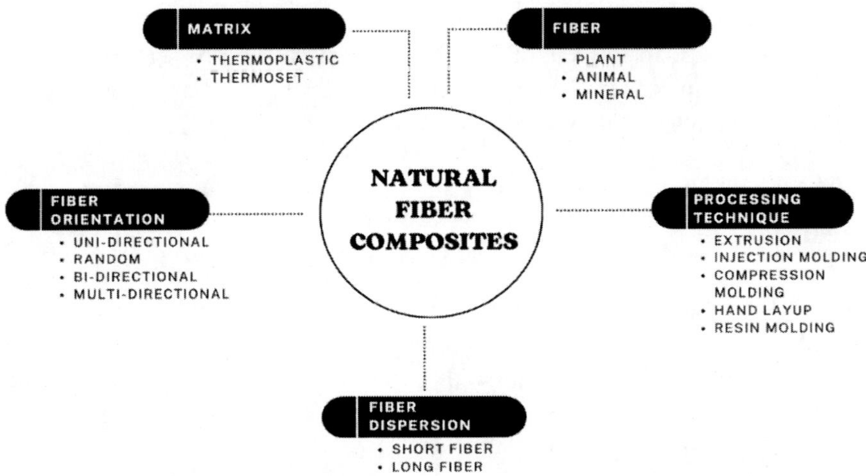

Figure 3. Factors affecting the features of NFC.

The mechanical properties of fiber-reinforced composites depend not only on the properties of the fiber but also on how much of the applied load the matrix phase can transfer to the fiber. The strength of the interfacial bond between the matrix and fiber phases determines how much load transmission occurs. Under loading conditions, the fiber-matrix bond terminates at the fiber ends. There is no load transmission from the matrix to the fiber ends.

Many factors must be taken into account when designing a fiber-reinforced composite.

Fiber length to diameter ratio: Continuous fiber production that gives the best properties is difficult. On the other hand, fibers with a large ratio of fiber length to fiber diameter provide high rigidity and strength.

Partial volume fraction of fibers: A higher amount of reinforcement can increase the strength and rigidity of the composite. On the other hand, there is an upper limit where the fibers cannot be wetted by the matrix. Fibers occupy nearly 30%–70% of the matrix volume in composites.

Orientation of fibers: Optimum rigidity and strength can be achieved if the load applied to the unidirectionally arranged fiber-reinforced material is parallel to the fibers. However, unidirectionally aligned fibers can show highly anisotropic behavior. If the fibers are arranged transversely and longitudinally or at an angle instead, maximum strength is compromised, but the composite gains more uniform properties.

Properties of fiber: An ideal fiber material should be strong, rigid, lightweight, and have a high melting temperature.

Matrix properties: Matrix materials are generally tough and ductile, transmitting the load to the fibers and preventing crack propagation through the composite in brittle fibers. The matrix must also be strong enough to contribute to the strength of the entire composite [13-16].

Treatment of Natural Fibers

For years, researchers have been studying how to enhance the mechanical properties of natural fiber-reinforced polymer composites by working with different fiber types, fiber ratios that should be employed, and surface treatments that should be given to natural fibers. The hydrophilic cellulose in the natural fiber and the hydrophobic nature of the synthetic matrix affect the interfacial bonding of the composite, in which case chemical treatment may be necessary. Surface treatments improve the interfacial bond between the matrix and fiber by increasing the surface roughness while reducing the number of hydroxyl functional groups on the fiber surface. Additionally, reducing particle size increases the consistency in the distribution of fibers and fillers within the biocomposite. Thus, mechanical properties are improved [7].

The hydroxyl (−OH) groups present in the structure of natural fibers cause it to react with water. This causes weak bonds at the composite interface. It negatively affects the properties of the composite due to the weak interfacial bond between hydrophilic fibers and hydrophilic polymers. Interface improvement has 4 main purposes. It is to reduce the polar component of natural fibers in order to eliminate impurities, change crystallinity, improve fiber-matrix interface bonding, and ensure good performance.

During surface treatments, chemicals react with functional groups on the natural fiber surface and provide surface modification. Thus, the interfacial interaction between the modified wood surface and the polymer matrix can be improved. More than 40 different chemical agents have been reported to be used to modify the surface of wood fibers. Among these, sodium hydroxide, acetic acid, silane, acrylic acid, peroxide, isocyanates, and potassium permanganate are the most common. Chemical reactions between natural fibers and the binder improve the compatibility and interfacial bonding in natural fiber-based polymer composites and reduce the water absorption of the fibers [9].

Chandekar et al. (2019) examined the effect of chemical treatments applied to jute fiber with alkali, potassium permanganate, and silane on the properties of the composite in a 30% jute fiber-added polypropylene composite. Chemically treated jute fibers were combined with 30 wt% PP granules in a twin-screw extruder, and then the pellets were injection molded to obtain composite samples. Potassium permanganate-treated jute-PP composites showed higher mechanical properties (tensile strength and elongation at break) among all other composites in the study [17].

In another study, Moczo et al. (2019) examined the properties of a sugarcane bagasse-reinforced polypropylene composite. In the study, a wood flour-filled composite was used as a reference. Maleic polypropylene was used as a coupling agent in both fillers, and it was evaluated that the coupling agent should be used to obtain good properties [18].

Surface treatments are grouped under three headings: biological, chemical, and physical. Biological enzyme treatment is widely used in the surface modification of natural fibers. Physical processes are also the preferred method for surface treatment. With these methods, the surface and structural properties of the fiber change, and the interface is modified. Some of the physical processes are thermal, plasma, hybridization, gamma, and ultraviolet surface treatments. Thermal treatment involves conditioning wood fibers with heat and moisture at approximately 230°C to eliminate hemicellulose. In the plasma process, different ionization gases are used to eliminate unwanted components from the surface of the fibers. With the hybridization process, hybridization can be achieved by combining different types of fibers or fillers in the same matrix to create a hybrid structure. High-energy gamma radiation can change the properties of polymer surfaces. With the UV radiation process, the polarity of the fibers changes, increasing interfacial adhesion and wettability [19].

Chemical treatments are also extensively used for surface modifications. Some of these are as follows:

- In the acetyl treatment process, cellulose fibers react with chemicals consisting of the acetyl (CH_3CO) group to improve the interfacial adhesion of natural fibers and lose the hydroxyl group.
- Alkaline treatment: In this method, sodium hydroxide reacts with the hydroxyl groups on the surface of natural fibers, and moisture absorption sensitivity is reduced by the hydroxyl groups.
- Benzyl treatment: With benzoyl chloride, it reduces the hydrophilicity of the natural fiber, improving the fiber-matrix interface adhesion and thermal stability of the composite.
- Isocyanate treatment: Bonds are formed when the functional groups of isocyanates react with the hydroxyl groups of lignin and cellulose.
- Permanganate treatment: In this process, the roughness of the fibers is improved. First, alkaline pretreatment is carried out, and then it is kept in a potassium permanganate solution containing acetone.
- Silane treatment is the process of dipping hydrophilic fibers into a silane solution in alcohol or acetone and bonding them with a hydrophobic structure.
- Maleated coupling agent treatment: Maleated coupling agents interact with hydroxyl groups during grafting and remove them from fiber cells. As a result, the long-chain polymer is covalently bonded to the fiber surface. Thus, the interfacial adhesion between the matrix and fiber is improved. Before the process, the copolymer is heated to 170°C, and then the esterification process is carried out [20].

Degradation

Both thermoplastic and thermoset-based resins are chosen in composites with natural fibers. However, in most applications, polyolefins are chosen. The use of polyolefins requires the processing temperature to be below 200°C. Above this temperature, degradation of fibers and fillers has been observed [21-22].

Richters and Moritzer (2021) examined the characteristics of a composite material with thermoplastic polyurethane matrix and wood fiber reinforcements. Different samples were obtained by changing the fiber type, fiber amount, and injection conditions. It has been said that strengthening properties can be achieved by increasing the amount of fiber. It was observed that the highest tensile strength was reached at a processing temperature of 190°C. Increasing the fiber amount caused an increase in hardness, but it was observed that fiber type did not affect hardness [23].

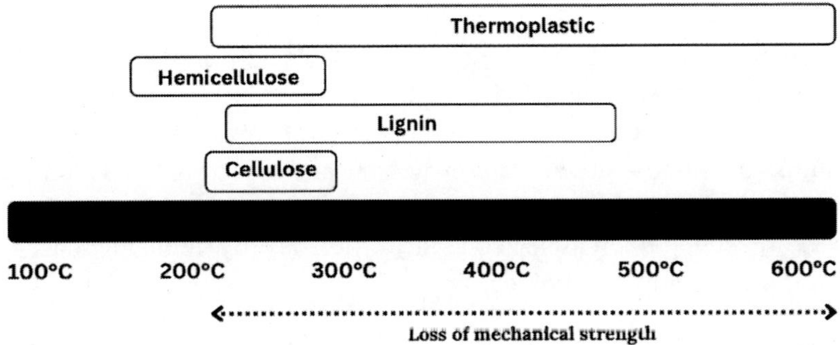

Figure 4. Degradation temperatures in natural fiber-based composites.

Degradation is the process of breaking down polymers into pieces of various structures and sizes. Degradation refers to the loss of the original properties of a polymeric material exposed to the environment and workload. Although degradation is generally an undesirable process, it can be beneficial in some cases. For example, it can improve the processing of polymers or be used in the natural degradation and recycling of waste polymers [24].

The degradation of polymers is examined in 4 groups: thermal degradation, photodegradation, catalytic degradation, and biodegradation. Generally, energy from heat and light can be absorbed by polymer chains, leading to thermal and light cracking, respectively. The presence of oxygen promotes these two reactions. Because of oxidation, various radical structures that can react with other chains are formed. The energy generated at the processing temperature of polymers encourages the movement of the polymer chain, causing polymer degradation, decomposition, and a decrease in mechanical properties. The chemical bonds absorb enough energy from the heat in the polymer, resulting in the breaking of the chemical bonds and the subsequent formation of some free radicals. The free radicals formed are extremely active and react with surrounding molecules. Thus, more chains can be broken, and more free radicals can be produced. In this way, polymer properties such as molecular weight, degree of crystallinity, and degree of branching will change. Therefore, the processing temperature of polymers containing natural fibers must be at a temperature that will not cause degradation of the fibers and fillers [7, 25].

Use of Recycled Polymers for NFC

We can use thermoplastics in a more wholesome and extended manner because of their recycling. Thermoplastics are used in daily life in the manufacture of products such as toys, kitchenware, decorative items, furniture, greenhouse covers, and irrigation hoses used in agricultural areas. Recycling plastics reduces the consumption of natural resources. In this respect, recycling increases economic efficiency by reducing energy and material use per unit output [26].

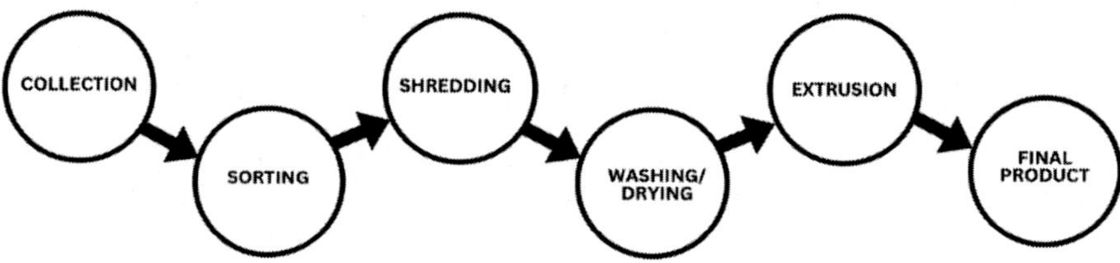

Figure 5. Recycling stages of plastics.

Recycling is the result of intermediate steps such as collection, sorting, and processing of polymers. In this way, the amount of waste in landfills is reduced. Thus, it prevents the pollution of soil, air, and water and prevents diseases that may occur [27].

There are concerns that the amount of plastic use has increased and will increase due to the increased personal plastic use after the COVID-19 pandemic [28]. One of the recycling methods for these plastics is their reuse after recycling. The use of recycled thermoplastics in natural fiber-reinforced composites is also common.

Studies have shown that the use of virgin or recycled matrix does not bring about a significant change in many properties. Singh et al. (2021) added 20% pineapple fiber by weight as a reinforcement to virgin HDPE and recycled HDPE, which they chose as matrix materials. It has been observed that pineapple fiber-filled composites have improvements in tensile strength compared to pure polymers, and whether the matrix used is virgin or recycled does not have much effect [29].

In another study, Lila et al. (2018) prepared bagasse fiber-reinforced polypropylene composite samples. They added 20% bagasse fiber by weight to polypropylene to obtain the composite material and then recycled the material 5 times. As a result of the study, it was observed that crushing the fibers during recycling increased the aspect ratio of the fibers and thus increased the tensile strength of the composite. The increase in crystallinity led to higher elongation under tensile loading, but the increase in surface roughness and hardness led to a decrease in bending properties. There was no significant change in thermal degradation behavior after 5 cycles. In the study, it was evaluated that recycled natural fiber-added composites could replace the use of traditional plastic in various applications [30].

Kartal and Selimoğlu (2023) examined the co-usability of wood sawdust and waste cotton in recycled polypropylene matrix composites. Wood sawdust is a product that emerges as waste after processing, especially in the furniture industry. Cotton, on the other hand, is used extensively in the textile industry and also occurs as a waste product in large quantities. Composite samples were obtained by adding wood sawdust in the 0-250 micron size range and waste cotton in the 1 cm size range to recycled polypropylene at different weight ratios. The samples were prepared by the injection molding method. The mechanical properties of the samples, such as tensile, impact, and hardness, were examined. It has been stated that the tensile strength, hardness and water absorption of composite samples increased compared to pure polypropylene, while impact resistance and MFR decreased. As a result of the study, it was reported that cotton and pine sawdust can be used together as fillers in polypropylene matrix composite materials, and the ideal ratio is 10% pine sawdust and 10% waste cotton [31].

Production Methods

Various processing methods are available for the production of fiber-reinforced composite materials. The appropriate processing method is selected depending on the desired product quality, feature, quantity, and cost. Some processing techniques are given below.

Compression molding is the most common in closed mold production, as it allows precise production of large quantities of parts. The basic principle of this process is based on heating the thermoplastic resin under pressure in a closed mold cavity until it cures. In this production, prepreg is placed in the heated mold cavity and compressed with a hydraulic press to ensure equal distribution throughout the mold.

Extrusion molding is another closed molding technique. It is a continuous process used in the production of products such as pipes. Fiber and matrix are added to the extruder from the feed hopper and moved forward with the help of the screw. In the meantime, the heaters surrounding the barrel start to soften and melt the mixture. Fiber-reinforced composite material is obtained by passing the molten mixture through the mold [32].

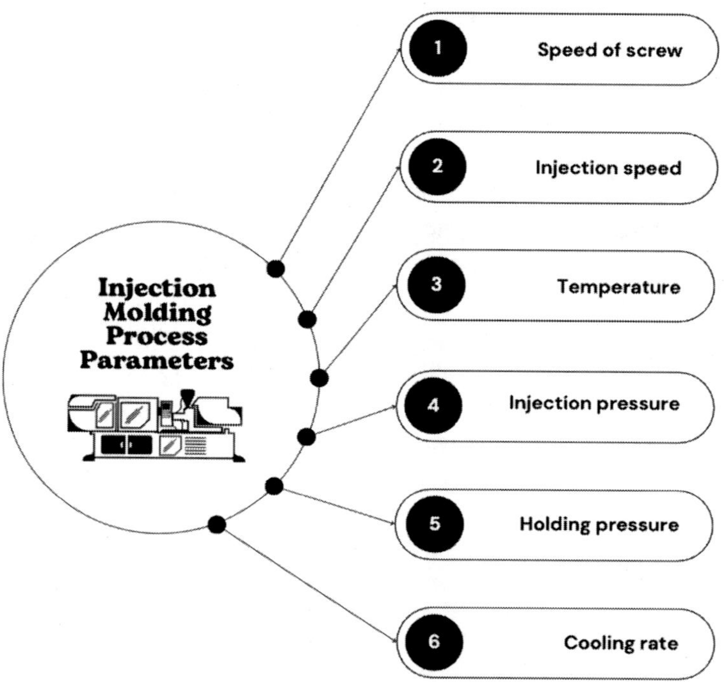

Figure 6. Process parameters during injection molding.

Injection molding technique can be applied to many polymer types and composites. In this technique, short-form fibers and particles are generally used as reinforcement. Mostly extrusion process is performed and then an injection molding process is carried out. First, polymers and short fibers are fed into the hopper of the extrusion machine. Then, as the polymer pellets and fibers move through the barrel, the materials begin to melt with heating. Heating is provided both by the heating elements and the cutting movement of the materials inside the barrel. In addition, the rotating movement of the screw facilitates the mixing of molten polymers and fibers. The molten mixture passes through the mold cavity and passes through the water bath to be cooled in wire form. After cooling, it is cut into granules and dried at the appropriate temperature before injection molding. These granules are used as raw materials and are placed in the feeding hopper of the injection machine. When the granules enter the barrel, the material softens with heat. As the screw rotates, premixed fibers and polymers move through the barrel. With the help of the nozzle at the end of the barrel, the material is injected into the mold cavity with high pressure. After curing is completed, the mold is opened and the part inside is removed with the help of ejector pins.

In direct injection molding, short fibers and polymer pellets are mixed by hand and added directly into the feed hopper of the injection molding machine. After melting and mixing, the mixture is injected into the mold cavity under high pressure. After cooling, the composite product is taken out. To avoid fiber damage due to the high temperature and shearing action, an additional hopper for adding natural fibers is sometimes placed near the injection end of the screw. Screw speed, injection pressure, injection speed, temperature, holding pressure, and cooling rate are important parameters during injection molding. With injection molding, parts of the desired size and shape can be produced with high accuracy [33].

Applications

Natural fiber-filled composite materials have recently begun to be used in many sectors. Especially considering the cost and environmental advantages, the use of natural fiber-filled composites instead of synthetic fibers is of great importance. For example, due to the non-abrasive nature of natural fibers, the product or material is safer during collision or impact compared to glass fiber. [34].

In the aerospace industry, many parts in the manufacturing of aircraft, helicopters, and drones are made of composite materials. Natural fiber-reinforced composites are also used in these parts. Especially due to the desire to reduce fuel consumption, natural fiber-reinforced composites, which are lighter than traditional materials, are preferred. Recently, natural fibers with less density have started to be used instead of glass fiber. In the automobile industry, many parts of vehicles have begun to be produced from natural fiber composite materials [35].

The first NFC studies in the automotive industry were in the 1930s when Henry Ford used hemp for automobile exteriors. In the following years, the use of natural fiber composite materials in the automobile industry has gradually increased. Green composites are achieving great success in the automotive industry with their environmental friendliness and superior lightness.

The luggage compartment of the Toyota CT 200 is made of bioplastics obtained from sugar cane, the Audi A2 door trim panels are made of flax/silicone-reinforced polyurethanes, and the spare tire well covers of the Mercedes-Benz A-Class coupe cars are made of abaca fiber-reinforced PP composites. Daimler-Benz used a variety of natural fibers (such as sisal, jute, coconut, European hemp, and flax) to replace glass fibers, making instrument panels, center armrest consoles, seat shells, and trim on seat backs, BMW used linen and sisal fibers for the car's interior door trim panels, uses cotton for sound insulation, wool for its upholstery and wood fiber for the seat backs. Recently, Daimler Chrysler cars have started to use banana fiber-reinforced polypropylene composites for underfloor protection [7, 36, 37].

Natural fiber reinforcements are also used in the construction industry, such as in concrete. Concrete is a source of carbon emissions; therefore, choosing ecologically friendly materials increases sustainability [38].

In the production of outdoor decoration and garden furniture, wood-polymer composite materials are preferred instead of traditional wooden materials due to their many superior properties. These materials are provided from waste wood sawdust from saws or waste wood products from old furniture and construction wood residues. In this group of composites, waste sawdust filler can be used in the range of 30-70%.

Wood products are often preferred as cheap fillers in PP and PVC composites. These composites containing wood fiber are used as outdoor floor coverings, window sills, and door sills. In these applications, since the composite material does not experience high stress, its mechanical properties do not need to be high. The use of recycled plastic for this group of composites is also very common [39].

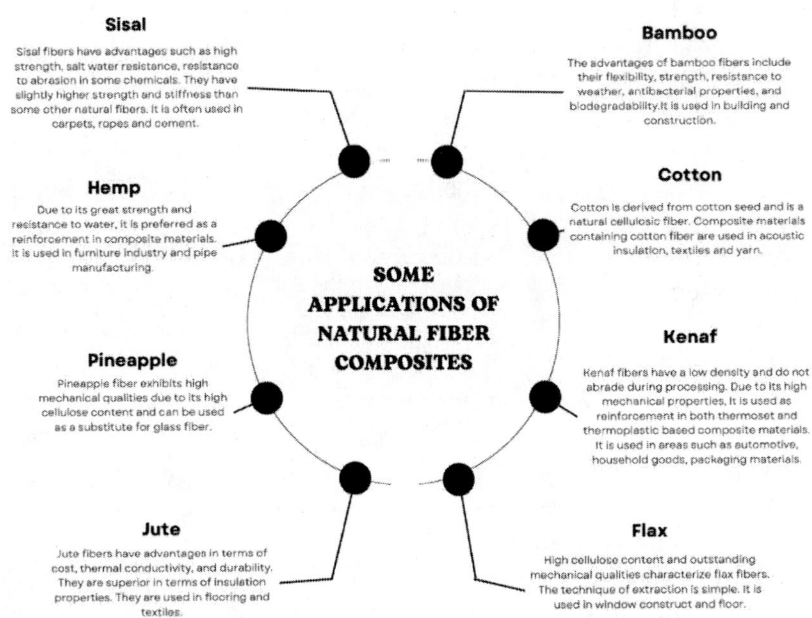

igure 7. Some applications in natural fiber-based composites.

100% bio-based coffins reinforced with natural fibers have also been preferred recently. It can be produced in various shapes with freedom of color and design. Bio-based pots produced from agricultural fibers obtained from pepper and tomato cultivation are used for tree nurseries and market gardens. Bamboo-based composite materials are also used in toys, food containers, and musical instruments. [40].

Conclusion

In polymer-based composites, the use of natural fibers or fillers is widely preferred in both thermosets and thermoplastics. Degradation of cellulose-based natural fibers occurs in thermoplastics due to the processing temperature. For this, it is important to control the processing temperature.

One of the factors affecting the final properties of composite materials is their interface properties. Interface properties can be improved by surface treatments between the fiber and the matrix. Thus, composite materials with more ideal properties can be obtained.

As a result, the use of natural fiber in polymer composites offers many advantages, such as sustainability, cost, and improvement of the properties of the composite. Improvements in the mechanical, thermal, and barrier properties of natural filler-reinforced composites, their diversity, and potential are becoming increasingly important. Despite challenges including agglomeration and homogenous dispersion, a great deal of work is taking place in order to create these materials.

As the industry seeks sustainable and environmentally friendly solutions, the use of natural fibers in polymer composites holds great promise for reducing our reliance on renewable resources and reducing the environmental impact of material production. As research continues, we can expect these composites to have wider and more diverse applications in the future. Using natural fibers will not only contribute to increased material performance but will also make a positive contribution to a more sustainable and environmentally friendly future.

Disclaimer

None

References

[1] Qiao Y, Fring LD, Pallaka MR, Simmons KL. 2023. "A Review of the Fabrication Methods and Mechanical Behavior of Continuous Thermoplastic Polymer Fiber–Thermoplastic Polymer Matrix Composites." *Polymer Composites* 44(2):694-733. DOI: 10.1002/pc.27139.

[2] Usman MA, Momohjimoh I, Usman AO. 2020. "Mechanical, Physical and Biodegradability Performances of Treated and Untreated Groundnut Shell Powder Recycled Polypropylene Composites." *Materials Research Express* 7(3):035302. https://doi.org/10.1088/2053-1591/ab750e.

[3] Jan P, Matkovič S, Bek M, Perše LS, Kalin M. 2023. "Tribological Behaviour of Green Wood-Based Unrecycled and Recycled Polypropylene Composites." *Wear* 524-525: 204826. https://doi.org/10.1016/j.wear.2023.204826.

[4] Shah AUR, Imdad A, Sadiq A, Malik RA, Alrobei H, Badruddin IA. 2023. "Mechanical, Thermal, and Fire Retardant Properties of Rice Husk Biochar Reinforced Recycled High-Density Polyethylene Composite Material." *Polymers* 15(8):1827. https://doi.org/10.3390/polym15081827.

[5] Theja KK, Bharathiraja G, Sakthi Murugan V, Muniappan A. 2023. "Evaluation of Mechanical Properties of Tea Dust Filler Reinforced Polymer Composite." *Materials Today: Proceedings* 80(3): 3208-3211. https://doi.org/10.1016/j.matpr.2021.07.213.

[6] Nneka Anosike-Francis E, Ijeoma Obianyo I, Wasiu Salami O, Odochi Ihekweme G, Ikpi Ofem M, Olajide Olorunnisola A, Peter Onwualu A. 2022. "Physical-Mechanical Properties of Wood Based Composite Reinforced with Recycled Polypropylene and Cowpea (Vigna unguiculata Walp.) Husk." *Cleaner Materials* 5:100101. https://doi.org/10.1016/j.clema.2022.100101.

[7] Dev B, Rahman A, Alam R, Repon R, Nawab Y. 2023. "Mapping The Progress in Natural Fiber Reinforced Composites: Preparation, Mechanical Properties, and Applications." *Polymer Composites* 44(7): 3748-3788. https://doi.org/10.1002/pc.27376.

[8] Kartal İ, Özcan Z. 2023. "Investigation of Effect of Chestnut Sawdust on Mechanical Properties of Epoxy Matrix Composites." *Journal of Innovative Engineering and Natural Science* 3(2): 67-74. DOI: 10.29228/ JIENS.69363.

[9] Khoaele KK, Gbadeyan OJ, Chunilall V, Sithole B. 2023. "A Review On Waste Wood Reinforced Polymer Composites And Their Processing For Construction Materials." *International Journal of Sustainable Engineering* 16(1):104-116. DOI: 10.1080/19397038.2023.2214162.

[10] Kartal İ, Naycı G, Demirer H. 2019. "Cam ve Bambu Lifleriyle Takviyelendirilmiş Vinilester Kompozitlerinin Mekanik Özelliklerinin İncelenmesi." *International Journal of Multidisciplinary Studies and Innovative Technologies* 3(1): 34-37.

[11] Kartal İ, Naycı G, Demirer H. 2020. "The Effect of Chestnut Wood Flour Size on the Mechanical Properties of Chestnut Wood Flour Filled Vinylester Composites." *Emerging Materials Research* 9: 1-6. https://doi.org/10.1680/jemmr.19.00179.

[12] Kartal İ, Selimoğlu H. 2023. "Usability of Pine Sawdust and Calcite Together as Filler in Polyester Composites." *International Journal of Computational and Experimental Science and Engineering* 9(3): 267-273. DOI: 10.22399/ijcesen.1335325.

[13] Balasubramanian, M. 2013. *Composite Materials and Processing*. Boca Raton: CRC Press.

[14] Chauhan AK, Singh A, Kumar D, Mishra K. 2021. "Properties of Composite Materials." In *Composite Materials: Properties, Characteri-sation, and Applications*, edited by Sachdeva A, Singh PK, Rhee HW, Boca Raton: CRC Press.

[15] Hsissou R, Seghiri R, Benzekri Z, Hilali M, Rafik M, Elharfi A. 2021. "Polymer Composite Materials: A Comprehensive Review." *Composite Structures* 262:113640 https://doi.org/10.1016/j.compstruct.2021.113640.

[16] Koppiahraj K, Bathrinath S, Narayanasamy P, Balasundar P, Parrthipan BK, Senthil S, Ramkumar T. 2021. "Chapter Ten- A Review on The Factors Influencing Natural Fiber Composite Materials." In *Hybrid Natural Fiber Composites*, edited by Khan A, Rangappa SM, Siengchin S, Jawaid M, Asiri AM., 185-205. Woodhead Publishing.

[17] Chandekar H, Chaudhari V, Waigaonkar S, Mascarenhas A. 2020. "Effect of Chemical Treatment on Mechanical Properties and Water Diffusion Characteristics of Jute-Polypropylene Composites." *Polymer Composites* 41(4):1447–1461. https://doi.org/10.1002/pc.25468.

[18] Anggono J, Farkas ÁE, Bartos A, Móczó J, Antoni, Purwaningsih H, Pukánszky B. 2019. "Deformation and Failure of Sugarcane Bagasse Reinforced PP." *European Polymer Journal* 112: 153-160. https://doi.org/10.1016/j.eurpolymj.2018.12.033.

[19] Nurazzi NM, Harussani MM, Aisyah HA, Ilyas RA, Norrrahim MNF, Khalina A, Abdullah N. 2021. "Treatments of Natural Fiber as Reinforcement in Polymer Composites-A Short Review." *Functional Composites and Structures* 3(2): 024002. DOI:10.1088/2631-6331/abff36.

[20] Ferreira DP, Cruz J, Fangueiro R. 2019. "Chapter 1- Surface modification of natural fibers in polymer composites." In *Green Composites for Automotive Applications*, edited by Koronis G, Silva A., 3-41. Woodhead Publishing.

[21] Singh P, Singari RM, Mishra RS, Bajwa GS. 2022. "A Review on Recent Development on Polymeric Hybrid Composite and Analysis of Their Enhanced Mechanical Performance." *Materials Today: Proceedings* 56: 3692-3701. https://doi.org/10.1016/j.matpr.2021.12.443.

[22] Albinante SR, Platenik G, Batista LN. 2017. *Handbook of Composites from Renewable Materials, Volume 4, Functionalization*. USA: John Wiley & Sons, Inc; 2017.

[23] Moritzer E, Richters M. 2021. "Injection Molding of Wood-Filled Thermoplastic Polyurethane." *Journal of Composites Science* 5(12):316. https://doi.org/10.3390/jcs5120316.

[24] Vohlidal J. 2021. "Polymer Degradation: A Short Review." *Chemistry Teacher International* 3(2):213-220. https://doi.org/10.3390/jcs5120316.

[25] He Y, Li H, Xiao X, Zhao X. 2021. "Polymer Degradation: Category, Mechanism and Development Prospect." *E3S Web Conference* 290:01012. https://doi.org/10.1515/cti-2020-0015.

[26] United States Environmental Protection Agency. 2023. "*National Overview: Facts and Figures on Materials, Wastes and Recycling*" https://www.epa.gov/facts-and-figures-about-materials-waste-and-recycling/national-overview-facts-and-figures-materials#NationalPicture.

[27] Subramanian MN. 2017. *Polymer Blends and Composites: Chemistry and Technology*. Scrivener Publishing LLC.

[28] Zhao X, Long Y, Xu S, Liu X, Chen L, Wang YZ. 2023. "Recovery of Epoxy Thermosets and Their Composites." *Materials Today* 64:72-97. https://doi.org/10.1016/j.mattod.2022.12.005.

[29] Singh Y, Kumar J, Naik TP, Pabla BS, Singh I. 2021. "Processing and Characterization of Pineapple Fiber Reinforced Recycled Polyethylene Composites." *Materials Today: Proceedings* 44(1):2153-2157. https://doi.org/10.1016/j.matpr.2020.12.278.

[30] Lila MK, Singhal A, Banwait SS, Singh I. 2018. "A Recyclability Study Of Bagasse Fiber Reinforced Polypropylene Composites." *Polymer Degradation and Stability* 152:272-279. https://doi.org/10.1016/j.polymdegradstab.2018.05.001.

[31] Selimoğlu, H. 2023. "*Atık Pamuk ve Çam Talaşı Dolgulu Geri Dönüşüm Polipropilen Esaslı Kompozit Malzemelerin Mekanik Davranışlarının İncelenmesi.*"/"*Investigation of Mechanical Behavior of Recycled Polypropylene Based Composite Materials Filled with Waste Cotton and Pine Sawdust.*" MSc diss., Marmara University.

[32] Maiti S, Islam MR, Uddin MA, Afroj S, Eichhorn SJ, Karim N. 2022. "Sustainable Fiber-Reinforced Composites: A Review." *Advanced Sustainable System* 6(11): 2200258. https://doi.org/10.1002/adsu.202200258.

[33] Rakesh PK, Singh I. 2019. *Processing of Green Composites*. Singapore: Springer.

[34] Awais H, Nawab Y, Amjad A, Anjang A, Akil HMD, Abidin MSZ. 2021. "Environmental Benign Natural Fibre Reinforced Thermoplastic Composites: A Review." *Composites Part C: Open Access* 4:100082. https://doi.org/10.1016/j.jcomc.2020.100082.

[35] Gomez-Campos A, Vialle C, Rouilly A, Hamelin L, Rogeon A, Hardy D, Sablayrolles C. 2021. "Natural Fibre Polymer Composites - A Game Changer for the Aviation Sector?" *Journal of Cleaner Production* 286:124986. https://doi.org/10.1016/j.jclepro.2020.124986.

[36] Mochane MJ, Magagula SI, Sefadi JS, Mokhena TC. 2021. "A Review on Green Composites Based on Natural Fiber-Reinforced Polybutylene Succinate (PBS)." *Polymers* 13(8):1200. https://doi.org/10.3390/polym13081200.

[37] Samal KS, Mohanty S, Nayak SK. 2009. "Banana/Glass Fiber-Reinforced Polypropylene Hybrid Composites: Fabrication and Performance Evaluation." *Polymer-Plastics Technology and Engineering* 48(4):397-414. DOI: 10.1080/03602550902725407.

[38] Hamada HM, Shi J, Al Jawahery MS, Majdi A, Yousif ST, Kaplan G. 2023. "Application of Natural Fibres in Cement Concrete: A Critical Review." *Materials*

Today Communications 35: 105833. https://doi.org/10.1016/j.mtcomm.2023.105833.

[39] Netravali AN, Chabba S. 2003. "Composites Get Greener." *Materials Today* 6(4): 22-29. https://doi.org/10.1016/S1369-7021(03)00427-9.

[40] Mathijsen D. 2016. "Composites Made From Recycled Thermoplastics and Natural Fibers: A New Philosophy for the Future." *Reinforced Plastics* 60(3):142-145. https://doi.org/10.1016/j.repl.2016.04.068.

Chapter 6

Injection Molding of Rubbers

Seda Bekin Acar*
Department of Polymer Materials Engineering, Faculty of Engineering, Yalova University, Yalova, Turkey

Abstract

Rubbers are elastomeric materials that can change shape reversibly under force and return to their original dimensions when the deforming force is removed. Rubber compounds can be prepared to provide a wide range of properties, from soft to hard, depending on the intended use. The manufacture of rubber products for latex and dry rubber forms is entirely different, but vulcanization and some processing steps are common for both materials. Injection molding is highly preferred for the mass production of various polymeric materials and has been used in related sectors for many years. This processing method is preferred for reasons such as versatile shapes, low cost, short cycle times, and simple automation. When the polymeric material is rubber, the material with the desired geometry is obtained through the rubber injection molding process by injecting the rubber mixture into a closed mold and then vulcanizing it. In this chapter, different rubber processing methods are explained. Also, the basic concept of rubber injection molding and the factors affecting this process are discussed with relevant literature.

Keywords: curing, elastomer, injection molding, processing, rubber, vulcanization

* Corresponding Author's Email: seda.acar@yalova.edu.tr.

In: Injection Molding
Editor: Jose D. Phillips
ISBN: 979-8-89113-429-4

Introduction

Elastomers are polymer materials that are viscoelastic in nature and include natural and synthetic rubbers. Rubbers acquire the optimum properties of engineering materials only in vulcanizate form, which is possible only by means of vulcanization (Thomas and Maria, 2016). The basic principle in the vulcanization process is the formation of chemical and physical crosslinks between rubber macromolecules, which results in the formation of a three-dimensional network and the acquisition of special qualities by the material. In this case, elastomers may undergo reversible deformation in response to external forces that cause deformation. The degree of deformation is determined by the composition and molecular weight of the deformed rubber as well as by external deformation conditions; under low stress, it can already achieve a 100% to 1000% deformation.

Additionally, other additives need to be added to the rubbers. The mechanical and thermal characteristics of these materials are lower. In order to solve this problem, rubber compounds are prepared by adding different materials to the rubber matrix to increase its functional qualities (Acar et al., 2021; Acar et al., 2023). Compounding is the science of choosing and combining additives and elastomers to create a homogeneous mixture that will eventually acquire the required chemical and physical properties for the final product (Thomas and Maria, 2016). The main purposes of rubber compounding are to achieve balance between properties and price, and to meet the final properties and better processing requirements (Açar et al., 2023).

Processing requirements include homogeneous mixing of compound components in the elastomeric matrix. Mixing performance affects the viscosity, scorch safety and molding properties of the rubber compound. The molding process includes shaping and vulcanization. The fundamental concept of rubber molding is to form a crosslinked rubber compound by applying pressure to an uncured rubber compound inside a heated metal mold. In this chapter, the basic principles of rubber injection molding are summarized by considering the studies in the literature on this subject.

Processing of Rubbers

Rubber processing consists of four main steps, which are mastication, mixing, shaping of the viscous mass, and curing. In mastication, molecules are broken down and flow becomes easier. Mixing is usually accomplished by adding other ingredients immediately after mastication. The viscous mass can be shaped by extrusion or molding. In curing (vulcanization), polymer molecules are bonded to each other, and the shape is fixed. The plastic properties of rubber are replaced by elastic properties.

The most widely used techniques for manufacturing of rubber compounds are calendering, latex dipping, extrusion, and molding. Molding consists of three main production processes: compression molding, transfer molding and injection molding. With compression molding, the oldest and cheapest method, a rubber compound is formed into a cavity (rubber blank) and the blank is placed in a mold cavity to be shaped later. The curing time is long because the heating is slow, ranging from three minutes for thin walls to several hours for thick walls. This method has some benefits, including the ability to use rubber compounds with high viscosities and poor flow characteristics as well as rubber compounds with large surface areas. On the other hand, the slow process is a disadvantage due to the low production rate.

In the transfer molding, developed to limit the disadvantages of compression molding, a blank is initially loaded into the chamber and then distributed among numerous cavities to begin the process. First of all, pre-heating is carried out to facilitate the flow of rubber through the channels and allow it to fill the mold cavities efficiently. This pre-heating also speeds up the curing process and reduces the curing time. The disadvantages of this method are that the molds are more complex and expensive.

Injection molding has press and injection units that act as two separate units with separate controls. In the press unit, the molds can be placed horizontally or vertically. There can also be a moving extruder unit in a pre-programmed model. These enable short injection processes with high amounts of pre-heating. As a result, it is no longer necessary to handle blanks, procedures may be automated, and challenging cavities and flow channels are simply filled due to injection molding process.

Injection Molding of Rubbers

Injection molding is one of the most widely used processing methods for the production of technical rubber parts (Stanek et al., 2014; Ramini and Agnelli, 2023). The quality of the final product is affected by many different parameters such as rubber properties, mold design, and the capacity of the injection molding machine. Thus, rubber injection molding is a simple but complex rubber processing method (Traintinger et al., 2021). In this method, the rubber mixture is injected into a closed mold where it is shaped according to the required geometry. The rubber mixture that completely fills the cavity is vulcanized. So, rubber temperature has a very significant impact on the end-product quality. If the rubber temperature increases during injection, which reduces the viscosity of the rubber compound, the vulcanization reaction may begin. When curing occurs while the rubber is flowing, the viscosity of the rubber compound increases and it stops flowing (Ramini and Agnelli, 2023).

Vulcanization is the process in which uncured rubber with plastic properties is transformed into an elastic material by the formation of chemical crosslinks between polymer chains. The formulation recipe as well as the time and temperature history to which the material in the mold is exposed during curing determine the cure degree of the resulting rubber product (Arrillaga et al., 2012).

Injection molding of rubber, which began in the early 1940s, is now used in the manufacture of a wide variety of industrial products. This process basically involves placing a rubber mixture inside an injection unit and then injecting the rubber into a closed mold that allows it to take the shape of the cavity (Arrillaga et al., 2009). Temperature gradients continue to exist in the rubber after the cavity is filled, resulting in temperature distributions throughout the rubber's bulk. The material in the end zone of the filling section is subjected to intense heating, which causes its temperature to rise during the mold filling phase. Convection causes the material to continue heating as the mold is heated to a high temperature, and when a specific critical temperature is reached, the material starts to cure. The curing system used in the rubber formulation recipe determines this temperature. The degree of curing depends on the material temperature when the mold is completely filled, the temperature of the mold cavity and the curing time of the material.

Injection molding enables the production of complicated microstructured parts in a single-step procedure (Hopmann et al., 2014). Injection molding of elastomers is particularly important for industries with large-scale production such as automotive industry because it enables continuous production of high-quality parts at affordable cost. In this process, vulcanization, which is the longest stage in the production cycle, takes place in a closed mold. Therefore, accurate evaluation of mold temperature and vulcanization time during the production process is very important for cost analysis and process optimization (Arrillaga et al., 2007; Ramorino et al., 2010).

Natural rubber (NR) is one of the most widely used rubbers to produce various rubber products via injection molding. Although natural rubber products have been successfully produced through the injection molding process for many years, scientific studies on this subject have only just begun to develop. Experimental and modeling studies of rubber injection molding are carried out together to enable basic characterization of rubber compounds and prediction of the curing state in molded rubber products (Isayev et al., 1998).

Among elastomers, injection molding of liquid silicone rubber (LSR) continues to steadily improve due to its ability to produce complex elastomeric components with relatively short cycle times. LSR is used in most of the studies on rubber injection molding in the literature. The conventional plastic injection molding process and the liquid silicone rubber injection molding process are comparable in general (Park et al., 2016). Injection molding of liquid silicone rubber is a promising technique in the production of complex parts due to its easy processing conditions (Kuo and Lin, 2019). This process of LSR is discussed in detail in the "Injection Molding of Liquid Silicone Rubber (LSR)" section.

Process Control for Optimal Injection Molding of Rubber Compounds

Rubber compounds are chemically active materials. Since the properties of rubber products depend on storage time and temperature, their production poses a great challenge for manufacturers (Fasching et al., 2016).

Rubber injection molding is a widely used processing method that provides the production of rubber products (Stanek et al., 2014). Factors such as rubber type, mold design, process conditions and capacity of the injection molding machine determine the quality of the final product (Ramini et al., 2022).

Rubber compounds are crosslinked after the injection stage in the injection molding process (Fasching et al., 2016). Processing of rubber by injection molding is a highly complex process due to the influence of many parameters such as pressure, shear rate and especially temperature. Control of mold temperature is easy using thermocouples or other in-mold sensors. However, it is very difficult to control the rubber temperature as it varies according to parameters such as injection pressure and speed as well as screw rotation speed. Rubber temperature is a very critical parameter for the stages of curing and filling. Vulcanization may begin during cavity filling when the temperature of the rubber rises too much. This can cause defects in molded parts. Conversely, if a vulcanization reaction occurs as the rubber flows, the viscosity increases dramatically, and the flow nearly stops. Thus, before the vulcanization reaction, the mold cavity needs to be fully filled with rubber (Proske and Bhogesra, 2016). In summary, an appropriate rubber temperature is necessary to ensure a compromise between scorch safety (following complete mold filling, curing begins) and vulcanization initiation with increased productivity (Proske and Bhogesra, 2016; Ramini and Agnelli, 2023). Other molding issues resulting from excessively high temperature of rubber include mold fouling and adhesion, which are mostly caused by the components of the rubber compound diffusing (Ramini and Agnelli, 2020).

Injection Molding of Liquid Silicone Rubber (LSR)

Liquid silicone rubber (LSR) has a short history compared to other elastomers. LSR systems were developed in the late 1970s but have grown significantly in popularity over the past 30 years as a better understanding of their advantages has increased.

LSR, accessible in a broad variety of hardness, has superior chemical and physical properties because of its stable Si-O backbone. Important characteristics such as good mechanical and electrical properties over a wide temperature range, biocompatibility, and resistance to ozone, UV radiation, chemicals and aging increase the potential use of LSR (Shit and Shah, 2013; Harkous et al., 2016; Hopmann and Röbig, 2017; Bont et al., 2021; Weißer et al., 2023).

Depending on exposure times, LSR materials can be used at temperatures between -40°C and 200°C. Over time, chemical changes occurring on the surface of LSR components decrease the effects of UV radiation. LSR materials with a wide range of properties can be produced by different processing methods. One of these processing methods is injection molding of liquid silicone rubber. While LSR injection molding can be similar to thermoplastic injection molding in many ways, it can also be quite different from this process.

LSR is fed into the relatively cold chamber of an injection molding machine and then sent to a heated mold to cure in the LSR injection molding. Hot and cold runner systems are used in molds. In hot runner systems, the fully cured runner is removed with the part. On the other hand, the LSR is kept cold to prevent the runner from curing in cold runner systems. In this case, there is no runner waste in the part. Liquid silicone rubber is fed into an injection molding machine utilizing a pump. Component A contains the polymer and component B contains the catalyst. If color or other additives are required, other color pigment/additives plungers are added to system. All components are pumped into a static mixer in LSR molding production. After that, the liquid mixture is injected into a heated sealed mold with cavities. The liquid mixture injected into a closed mold is then heated within the mold to initiate curing. Once curing is complete, the molding machine ejects almost the final part. The vulcanization rate in this process depends on the geometry of the part, the temperature of the mold and possible inserts, the temperature of the silicone when it reaches the cavity, and the chemistry of the cure (Bont et al., 2021).

LSR composition, curing rate and time, part geometry and thickness, and mold temperature affect the injection rate. This injection rate is maximized to prevent premature curing, which causes surface defects, and to shorten cycle time. However, flashing from excessive pressure, air trapping, and severe material shearing that can separate fillers and color/additives limit the injection rate. Many commercial LSR injection molding manuals recommend completing the injection process in 0.5 to 3 seconds (Bont et al., 2021).

The LSR injection molding process offers the opportunity to produce high-quality rubber parts cost-effectively because of its high scalability (Haberstroh et al., 2002; Weißer et al., 2023). LSR parts have specific properties and good end product performance. Therefore, these parts are widely used in automotive, electrical, aerospace, medical, and consumer products (Hopmann et al., 2014; Kuo and Lin, 2019). Silicone is a very important material to provide engine insulation in the automotive industry due to its high decomposition temperature. In addition, it plays an important role in medical fields thanks to its biological inertness. In recent years, the drop in the raw material cost of LSR has increased their use and expanded their market share, making them competitive with cheaper thermoplastic elastomers. Complex LSR parts to be used in required areas are produced in very short production times by injection molding of silicone rubber. The complexity of the geometry plays an important role in determining the size of the mold and thus the molding machine.

In LSR injection molding, variables such as mold temperature, injection pressure, cure time, curing degree, and viscosity affect the process. In studies conducted at mold temperatures between 80°C and 220°C, it was observed that the curing rate, hardness, and shrinkage increased, while tensile strength and elongation at break decreased as the mold temperature increased. Also, viscosity decreased with vulcanization (Barbaroux et al., 1997; Haberstroh et al., 2002; Patel, 2002; Zhang et al., 2015). A change in shrinkage behavior was observed as the injection pressure changed (Patel, 2002; Matysiak et al., 2010; Zhang et al., 2015). The degree of curing increased with increasing curing time and temperature. Additionally, as the degree of curing increased and the temperature decreased, the viscosity also increased (Harkous et al., 2016). Part removal was possible when 75% to 95% of curing was complete (Haberstroh et al., 2002).

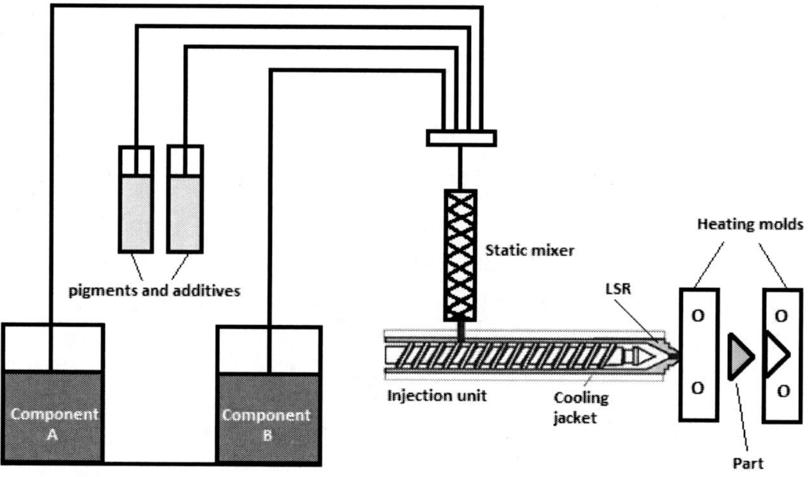

Figure 1. General injection molding process of liquid silicone rubber.

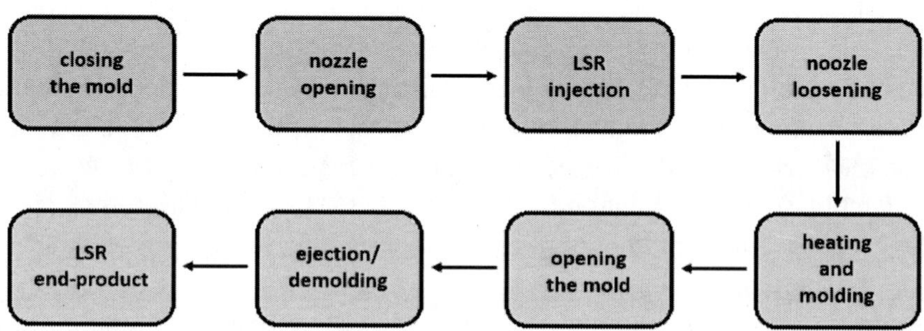

Figure 2. Injection process cycle of LSR.

Two Component Injection Molding: Elastomers and Thermoplastics

Two-component injection molding is a processing method generally used to combine polymers with certain properties without assembly. Recently rubbers and thermoplastics have been combined with a new two-component injection molding technique. This technique becomes important when thermoplastic parts also include parts made of rubber. Normally, in this method, hard thermoplastic material and rubber-like thermoplastic elastomer (TPE) are used as different components. However, when resistance to factors such as mechanical wear, chemical environment or high temperature is required, thermoset rubber is used instead of TPE (Bex et al., 2018).

The technical problem with two-component injection molding is that different conditions are required for the processing of thermoplastics and thermoset rubbers. In the injection molding process, thermoset rubbers are injected at low temperature and then cured in a heated mold, whereas thermoplastics are injected at high temperature and solidification occurs in a cooled mold (Lindsay, 2012; Bex et al., 2017; Bex et al., 2018). For this reason, a system has been developed that allows thermoplastic and thermoset rubbers to be produced in the same injection molding machine. This system has a mold in which the cavities of the thermoplastic part and the rubber part are thermally separated by an insulation plate and air gaps (Bex et al., 2017). The temperature control of both cavities is separate. Thus, during cooling of the thermoplastic part, the rubber is vulcanized simultaneously without any deformation (Bex et al., 2018).

Some Literature Studies on Injection Molding of Rubbers

In the literature, works on injection molding of rubbers generally include automated/monitoring systems and process modeling/simulation studies (Isayev et al., 1998; Haberstroh et al., 2002; Arrillaga et al., 2007; Ramorino et al., 2010; Stanek et al., 2011; Arrillaga et al., 2012; Rubenstein et al., 2018; Bont et al., 2021; Hutterer et al., 2021; Sanchez-Castillo et al., 2021; Traintinger et al., 2021; Ramini and Agnelli, 2023).

Table 1. Some Literature Studies on Rubber Injection Molding

Rubber Type	Purpose of the study	Results	Reference
LSR	To develop a cost-effective method for the production of polymer parts	- Improved mechanical properties - Increased heating rate	(Kuo and Lin, 2019)
Ethylene acrylate rubber	Process optimization considering scorch and thermal degradation	- High-speed online process control - Information on the behavior of rubber and the shear heating effect before injection - Calculation of the shear heating parameter	(Ramini et al., 2022)
Nitrile butadiene rubber	Change of rubber processing behavior due to material storage	- Restricted processability of rubber compounds - Decreased mechanical properties - Shrinkage behavior changing with storage	(Fasching et al., 2016)
Natural rubber	Simulation and experimental studies to comprehend processing-property relationships	- Consistent experimental and theoretical results - Two factors to consider: additional pressure losses and flow-induced crystallization phenomena	(Isayev et al., 1998)
LSR	Production of superhydrophobic functional surfaces	- Strong surface properties - Self-cleaning characteristics - Good molding quality	(Hopmann et al., 2014)
LSR	Cavity pressure analysis to investigate crosslinking of rubber	- A test device developed with pressure and temperature sensors - Evaluation of cross-linking via pressure curves over temperature or time - Opportunity for process monitoring and control in LSR molding	(Weißer et al., 2023)

Rubber Type	Purpose of the study	Results	Reference
Ethylene propylene diene monomer (EPDM)	The combination of EPDM with high density polyethylene (HDPE) and process parameters affecting the adhesion force between these two components	- Mold temperature: the most effective process parameter in adhesion strength - Holding pressure: only important if the holding pressure or holding time value is set too low - Other parameters (injection temperature, injection speed and interface roughness): no critical influence on adhesion strength	(Bex et al., 2018)
LSR	Filling and curing simulation	- Process setting provided for the injection stage - Possibility of calculation heating time	(Haberstroh et al., 2002)

Kuo and Lin developed a cost-effective method consisting of aluminum powder and epoxy resin for preparing polymer parts by injection molding of liquid silicone rubber. It was observed that the mechanical properties of the test samples prepared with commercially available materials improved significantly (Kuo and Lin, 2019).

In another study, Ramini and coworkers investigated the process optimization of ethylene acrylate rubber by injection molding, considering scorch and thermal degradation. They provided a process control system with online monitoring for the shear heating parameter (Ramini et al., 2022).

Fasching et al., studied the storage-induced changes in the processing behavior of injection molding of rubbers. Storage limited the processability of rubber compounds and significantly affected the mechanical properties and shrinkage behavior (Fasching et al., 2016).

In another study, both simulation and experimental studies were conducted for injection molding of natural rubber, considering previous studies for styrene butadiene rubber. It was observed that the experimental and theoretical results were in agreement with each other in terms of pressure during cavity filling and curing in the mold. However, it was concluded that extra pressure losses and flow-related crystallization events should also be taken into account when simulating rubber injection molding (Isayev et al., 1998).

Hopmann and coworkers aimed to produce superhydrophobic surfaces by injection molding of liquid silicone rubber. It was emphasized that with this method, self-cleaning and strong surface properties could be achieved due to the strong hydrophobic material (Hopmann et al., 2014).

Cavity pressure, a process parameter during injection molding of LSR, was also examined. A test device with pressure and temperature sensors was developed and pressure measurements were used to determine the crosslinking behavior of LSR. This study demonstrated that crosslinking characterization was possible by evaluating pressure curves over temperature or time (Weißer et al., 2023).

Some literature studies on rubber injection molding are summarized in Table 1.

Conclusion

In conclusion, rubber injection molding enables the continuous production of a wide range of complex rubber products in relatively low cycle times and at achievable cost. It becomes more important especially in sectors where high-scale production is carried out, such as the automotive industry. It is also used in fields of aerospace, medical and consumer goods. Although it is a method applicable to natural and synthetic rubbers, injection molding of liquid silicone rubber is highly preferred because it is comparable to the traditional plastic injection molding process.

Rubber injection molding also offers process control with simulation and monitoring systems in addition to experimental studies.

As a result, the injection molding method used in the processing of rubbers provides the discovery of new possibilities that can be achieved with rubber materials due to the properties mentioned in this chapter.

Disclaimer

None.

References

Acar SB, Tasdelen MA, Karaagac B. Methacrylate-functionalized POSS influence on cross-linking and mechanical properties of styrene-butadiene rubber. *Iranian Polymer Journal* (2021) 30(7):697-705.

Acar SB, Tasdelen MA, Karaagac B. The effect of POSS nanoparticles on crosslinking of styrene-butadiene rubber nanocomposites. *Turkish Journal of Chemistry* (2023) 47(2):417-425.

Açar SB, Tasdelen MA, Karaagac B. Haloysit içeren stiren-bütadien kauçuk nanokompozitlerinin hazırlanması ve mekanik özelliklerinin incelenmesi. *Journal of Innovative Engineering and Natural Science* (2023) 3:89-102.

Arrillaga A, Zaldua AM, Atxurra RM, Farid AS. Die pressure losses during injection molding of rubber mixes. *Rubber Chemistry and Technology* (2009) 82(1):62-93.

Arrillaga A, Zaldua AM, Atxurra R, Farid AS. Techniques used for determining cure kinetics of rubber compounds. *European Polymer Journal* (2007) 43(11):4783-4799.

Arrillaga A, Zaldua AM, Farid AS. Evaluation of injection-molding simulation tools to model the cure kinetics of rubbers. *Journal of Applied Polymer Science* (2012) 123(3):1437-1454.

Barbaroux M, Stalet G, Regnier G, Trotignon JP. Determination of the inter-relationships between processing conditions and properties of an injection molded silicone ring using an experimental design. *International Polymer Processing* (1997) 12(2):174-181.

Bex GJ, Desplentere F, De Keyzer J, Van Bael A. Two-component injection moulding of thermoset rubber in combination with thermoplastics by thermally separated mould cavities and rapid heat cycling. *The International Journal of Advanced Manufacturing Technology* (2017) 92:2599-2607.

Bex GJ, Six W, Laing B, De Keyzer J, Desplentere F, Van Bael A. Effect of process parameters on the adhesion strength in two-component injection molding of thermoset rubbers and thermoplastics. *Journal of applied polymer science* (2018) 135(29): 46495.

Bex GJ, Six W, Laing B, De Keyzer J, Desplentere F, Van Bael A. J. J. o. A. P. S. (2018b). Effect of process parameters on the adhesion strength in two-component injection molding of thermoset rubbers and thermoplastics. *Journal of Applied Polymer Science* 135(29):46495.

Bont M, Barry C, Johnston S. A review of liquid silicone rubber injection molding: Process variables and process modeling. *Polymer Engineering & Science* (2021) 61(2):331-347.

Fasching M, Friesenbichler W, Berger G. *Change of processing behavior of rubbers in injection molding caused by material storage.* Paper presented at the AIP Conference Proceedings (2016).

Haberstroh E, Michaeli W, Henze E. Simulation of the filling and curing phase in injection molding of liquid silicone rubber (LSR). *Journal of reinforced plastics and composites* (2002) 21(5):461-471.

Harkous A, Colomines G, Leroy E, Mousseau P, Deterre R. The kinetic behavior of liquid silicone rubber: a comparison between thermal and rheological approaches based on gel point determination. *Reactive and Functional Polymers* (2016) 101:20-27.

Hopmann C, Behmenburg C, Recht U, Zeuner K. Injection molding of superhydrophobic liquid silicone rubber surfaces. *Silicon* (2014) 6(1):35-43.

Hopmann C, Röbig M. Injection moulding of high precision optics for light-emitting diodes made of liquid silicone rubber. *Journal of Elastomers & Plastics* (2017) 49(1):62-76.

Hutterer T, Berger-Weber GR, Kerschbaumer RC, Friesenbichler W. J. P. E., & Science. Rubber injection molding: Applying multivariate statistics to identify quality issues solely from process signals. *Polymer Engineering & Science* (2021) 61(4):983-992.

Isayev AI, Wan M. Injection molding of a natural rubber compound: simulation and experimental studies. Rubber chemistry and technology (1998) 71(5):1059-1072.

Kuo CC, Lin JX. A cost-effective method for rapid manufacturing polymer rapid tools used for liquid silicone rubber injection molding. *The International Journal of Advanced Manufacturing Technology* (2019) 104:1159-1170.

Lindsay JA. *Practical guide to rubber injection moulding*: Smithers Rapra (2012).

Park H, Cha B, Cho S, Kim D, Choi JH, Pyo BG, Rhee B. A study on the estimation of plastic deformation of metal insert parts in multi-cavity injection molding by injection-structural coupled analysis. *The International Journal of Advanced Manufacturing Technology* (2016) 83:2057-2069.

Patel NB. *The effects of process conditions on the properties of injection molded liquid silicone rubber*: University of Massachusetts Lowell, ProQuest Dissertations Publishing (2002).

Proske M, Bhogesra H. Evaluating the root causes of rubber molding defects through virtual molding. *Rubber World* (2016) 255(3):20-26.

Ramini M, Agnelli S. Shear heating parameter of rubber compounds useful for process control in injection molding machine. *Rubber chemistry and technology* (2020) 93(4):729-737.

Ramini M, Agnelli S, Ramorino G. Applications of shear heating parameter for injection molding process optimization of AEM rubber compounds. *Express Polymer Letters* (2022) 16(4):354-367.

Ramini M, Agnelli S. Monitoring of shear heating effects during injection molding of rubber to improve the process control. *Polymer Bulletin* (2023) 80(6):6707-6723.

Ramorino G, Girardi M, Agnelli S, Franceschini A, Baldi F, Vigano F, Ricco T. Injection molding of engineering rubber components: a comparison between experimental results and numerical simulation. *International Journal of Material Forming* (2010) 3:551-554.

Rubenstein M, Handaya D, Slamet W. *Design and implementation of productivity monitoring system in rubber injection molding machine in real time based on visual studio and android application.* Paper presented at the IOP Conference Series: Materials Science and Engineering (2018).

Sanchez-Castillo L, Nedelcu D, Francisco-Marquez M. Redesign of layout runner in rubber injection molding for filling of a multi-cavity mold. *Materiale Plastice* (2021) 58(3):121-128.

Shit SC, Shah P. A review on silicone rubber. *National academy science letters* (2013) 36(4):355-365.

Stanek M, Manas D, Manas M, Javorik J. Simulation of injection molding process by cadmould rubber. *International Journal of Mathematics and Computers in Simulation* (2011) 5(5):422-429.

Stanek M, Manas D, Manas M, Ovsik M, Senkerik V, Skrobak A. Injection molding of rubber compound influenced by injection mold surface roughness. *Advanced Materials Research* (2014) 1025:283-287.

Thomas S, Maria HJ. *Progress in rubber nanocomposites*: Woodhead Publishing (2016).

Traintinger M, Kerschbaumer RC, Lechner B, Friesenbichler W, Lucyshyn T. Temperature profile in rubber injection molding: application of a recently developed testing method to improve the process simulation and calculation of curing kinetics. *Polymers* (2021) 13(3):380.

Weißer DF, Mayer D, Schmid J, Deckert MH. Novel approach to characterize the cross-linking effect of liquid silicone rubber via cavity pressure analysis during injection molding. *Journal of Applied Polymer Science* (2023) 140(4):e53381.

Zhang Y, Mischkot M, Hansen HN, Hansen P. *Replication of microstructures on three-dimensional geometries by injection moulding of liquid silicone rubber.* Paper presented at the Proceedings of the 15th International Conference on Metrology and Properties of Engineering Surfaces, ASPE, (2015).

Biographical Sketch

Seda Bekin Acar

Affiliation: Yalova University.

Education:
BS: Istanbul University, Department of Chemical Engineering.
MS: Istanbul University, Department of Chemical Engineering.
PhD: Yalova University, Department of Polymer Materials Engineering.

Business Address: Department of Polymer Materials Engineering, Faculty of Engineering, Yalova University, Yalova, Turkey.

Research and Professional Experience: Assistant Professor at Yalova University Polymer Materials Engineering Department since 2014.

Publications from the Last 3 Years:

[1] Turp, O., Acar, S. B., Ozdemir, K., Bouharras, F. E., Raihane, M., & Tasdelen, M. A. Halloysite Containing Thermoset Nanocomposites via Free Radical Photocrosslinking Polymerization. *Macromolecular Chemistry and Physics*, 2020. 221(21): p. 2000197.

[2] Acar, S. B., R. Ozdogan, and M. A. Tasdelen, POSS-based hybrid nanocomposites, in Polyhedral Oligomeric Silsesquioxane (POSS) *Polymer Nanocomposites*. 2021, Elsevier. p. 205-216.

[3] Acar, S. B., Ciftci, M., Bouharras, F. E., Raihane, M., & Tasdelen, M. A. In-situ preparation of halloysite nanotube-epoxy thermoset nanocomposites via light-induced cationic polymerization. *European Polymer Journal*, 2021. 158: p. 110682.

[4] Acar, S. B., M. A. Tasdelen, and B. Karaagac, Methacrylate-functionalized POSS influence on cross-linking and mechanical properties of styrene-butadiene rubber. *Iranian Polymer Journal*, 2021. 30(7): p. 697-705.

[5] Acar, S. B., Ozdemir, K., Raihane, M., Lahcini, M., & Tasdelen, M. A. Thermoset nanocomposites reinforced by vinyl-functionalized halloysite. *Polymer Composites*, 2023. 44(1): p. 148-155.

[6] Acar, S. B., M. A. Taşdelen, and B. Karaağaç, The effect of POSS nanoparticles on crosslinking of styrene-butadiene rubber nanocomposites. *Turkish Journal of Chemistry*, 2023. 47(2): p. 417-425.

[7] Acar, S. B., M. A. Tasdelen, and B. Karaağaç, Haloysit içeren stiren-bütadien kauçuk nanokompozitlerinin hazırlanması ve mekanik özelliklerinin incelenmesi. *Journal of Innovative Engineering and Natural Science*, 2023. 3(2): p. 89-102.

Index